THE McDONNELL DOUGLAS F/A-18 *HORNET* STORY

The prototype F/A-18A, BuNo. 160775, upon roll-out at the St. Louis, Missouri, McDonnell Douglas facility on September 13, 1978. The aircraft was painted white with blue and gold trim and had Navy markings on its port side and Marine Corps markings on its starboard. The nose anti-glare shield, ahead of the windscreen, was flat black. This aircraft apparently was the only single-seat "Hornet" to bear the hornet caricature on its nose; it later would have the number "1" painted in black in the white area of each vertical fin.

CREDITS:

The author and **Aerofax, Inc.** would like to express their sincere gratitude to John Brooks, Lon Nordeen, and Tom Downey of McDonnell Douglas Corporation, without whose assistance this **Minigraph** would not have been possible. Special thanks also goes to E. S. "Mule" Holmberg and Hughes Aircraft Co. in consideration of their never-ending research efforts on our behalf. Additional thanks go to the Canadian Armed Forces, George Cockle, Jim Goodall, Michael Grove, Henry Ham, Tom Hull, III, Lori Jordan, Gayle Lawson, Robert Lawson, Loral Systems Group public relations person Dorine Goss, Don McGarry, Northrop Corp. public relations personnel Terry Clawson, Tony Contafio and Suzanne O'Bourke, Tom Ring, Capt. Brian Rogers, Mick Roth, Keith Snyder, Bruce Trombecky, Mike Wagnon, and Barbara Wasson.

PROGRAM HISTORY:

The origins of McDonnell Douglas' F/A-18 *Hornet* can be traced directly to the birth of the Air Force lightweight fighter (LWF) program of the mid-1960s. The latter came to life during the early days of U.S. involvement in the Vietnam war when a rapidly growing data base began to reveal that extant U.S. fighters were generating a decidedly poor kill/loss ratio when pitted against their North Vietnamese adversaries. The cause quickly was traced to the inherent limitations imposed by the poor maneuvering performance, excessive size, and inadequate thrust-to-weight ratios of U.S. fighters. Aircraft such as the McDonnell F-4C proved a decidedly poor match when working one-on-one close-in with then-contemporary Soviet fighters such as the MiG-19 and MiG-21. Superior piloting skills rather than superior hardware, in fact, often proved the most important difference in determining the victor during engagements with N. Vietnamese-operated versions of these Soviet-designed aircraft.

Under the influence of such far-sighted military thinkers as John Boyd, Pierre Sprey, Chuck Meyers, and Everest Riccioni (who, along with several other visionaries of the day, became known collectively as the "Fighter Mafia") LWF philosophy had begun to evolve within select offices of the Department of Defense (DoD) when it was realized that simplicity and maneuverability could give a combatant a significant edge in an aerial dogfight. These fundamental precepts then were translated into a hardware design philosophy that predetermined a small airframe with minimal armament and minimal peripheral subsystems. Full emphasis was placed on maneuverability performance, pilot/cockpit interface, "visual range" engagement, systems simplicity, and performance in the most often utilized part of the dog fighting envelope—those speeds and related altitudes between .75 Mach and 1.5 Mach and sea level to 25,000 ft., respectively.

The resulting concept aircraft not only met the criteria for LWF design, it also had the attributes of costing less than half what larger and more complex fighters would cost—while proving to be a significantly more evasive and difficult-to-spot target.

Because of its size and associated simplicity, it was proposed that the LWF not be equipped with a powerful radar, that its armament system be absolutely limited to a pair of *Sidewinder*-type air-to-air missiles and an M61-type rotary gun, and that its maximum speed be Mach 2 or less.

When the LWF study was completed, it was presented to the AF for review. And as could be expected, the entrenched AF bureaucracy of the mid-1960s reacted adversely. Immediate concerns centered around the rapidly expanding F-X (Fighter-Experimental) program that was expected to result in the AF's next generation fighter (the F-15). AF planners felt that any other fighter, no matter how modest in scope, should be eliminated at conception to prevent any potential incursion into F-X funding.

In a preliminary effort to nip the LWF project in the bud, opponents made the fatal error of building opposing arguments around the very logic that was used to conceive the new fighter philosophy in the first place. They stated unequivocally that the LWF would have limited use in air combat scenarios due to (1) the limited range of its radar (dictated by nose radome cross-sectional area), (2) its minimal armament package, (3) its supposed short range, and (4) its limited maximum speed capability.

While LWF proponents began an on-going battle with the significantly larger F-X contingent, it began to dawn on the various major U.S. military aircraft manufacturers that potentially disastrous economic repercussions could result from a single-type production program as large as that planned for the F-X. The new fighter potentially could monopolize fighter funding for as long as two decades—thus leaving the losing bidders in a very tenuous financial position. Accordingly, motivation now was provided to force the DoD to fund *both* the F-X and the LWF—and effectively spread military appropriations over a broader segment of industry.

Virtually every major aircraft manufacturer now entered the fray. Included were Boeing, General Dynamics, Lockheed, McDonnell, Northrop, and Vought. Lockheed was quick off the mark with a highly modified F-104 design eventually referred to as the CL-1200 (and optimized as a replacement for NATO's expansive Lockheed F-104 fleet). After pushing it hard as a less-expensive F-X alternative (a fatal mistake, as the AF already had declared war on programs that threatened F-X funding), Lockheed conceded its failings and rapidly placed it on the back burner while quietly moving on to more politically palatable projects.

Northrop, while CL-1200 maneuvering was reaching its zenith, also initiated a DoD blitz in support of a design it had generated. For several years the company, under

The prototype Northrop YF-17, 72-1569, during a late 1974 test hop near Edwards AFB, California. The aircraft flew for the first time from Edwards AFB on June 9, 1974. Bearing a two-tone gray camouflage, it presently is preserved and awaiting disposition at Northrop's Hawthorne, California production facility.

1

F/A-18 production, in order to meet Navy and foreign inventory requirements, is maintained at a relatively high level at the McDonnell Douglas St. Louis facility. An F/A-18C is seen prior to painting as it nears final assembly before roll-out and the initiation of flight test. Its Hughes AN/APG-65 radar has yet to be installed.

the auspices of chief project engineer Lee Begin, had worked quietly on a series of design studies optimized, like Lockheed's CL-1200, to serve as a potential replacement for the NATO Lockheed F-104 fleet. Northrop estimated the total NATO market at some 3,000 fighters, with approximately 1,000 of these being susceptible to replacement with a Northrop product.

On January 31, 1971, Thomas Jones, then Chairman and President of Northrop, in a letter to the AF Chief of Staff, Gen. John Ryan, offered to build and flight test at company expense two pre-production P-530 fighters. Jones' letter started by specifically describing the P-530-3 proposal and then noting that as good as its predicted performance was in a heavily loaded condition (i.e., with a full avionics complement), performance could be greatly enhanced by the removal of all systems not essential to a single-purpose air-to-air combat fighter variant. The letter concluded by offering two austere P-530s to the AF for $15-million if the Dutch (who then were considering seriously the aircraft as a Lockheed F-104 replacement) would make the $100-million commitment required to initiate production. Later, an Air Staff assessment of the Northrop proposal determined it to be roughly similar to Lockheed's, with the exception of Northrop's superior corporate financial position and the required longer time from program go-ahead to first flight (27 months v/s 12).

The situation became somewhat more confusing on March 16, 1971, when Jones again wrote Gen. Ryan offering a further modified P-530, known as the P-530-4. This aircraft, referred to as an Air Research Vehicle (ARV), was proposed for use as an advanced aerodynamic and propulsion systems testbed. Importantly, initiation of this project no longer was contingent upon a foreign financial commitment.

The AF reacted to the newest Northrop proposal as they had the previous ones—with reservations. The company then was advised, on March 24, that no positive action would be taken on the P-530-4/ARV because of budgetary constraints, but that it would be reviewed again along with other similar proposals at a later date.

The initial rejection of the P-530 was a blow to Northrop, particularly in light of the fact that articles in both the U.S. and European aviation press then were stating (inaccurately) the AF was about to fund the Lockheed CL-1200. Some consolation later was garnered from the fact the AF subsequently elected to fund the General Electric GE-15 (later redesignated YJ101). This afterburning turbojet engine now was tied firmly to the P-530.

By the end of April 1971, it was clear that for either the Northrop or the Lockheed proposals to become serious contenders for funding, they would have to become part of larger prototyping programs. This, in turn, would have to be supported by Congress.

Northrop now made another significant move by proposing a program for a single-engine design, the P-610. This aircraft was based on the original P-530, but was considerably smaller and weighed only 17,000 lbs. (some 7,000 lbs. less than the P-530-4). Although a preliminary design and needing refinement, the P-610, with its lighter weight and smaller profile, was expected to have a better maneuverability envelope than its predecessor, particularly at high speeds.

While the various aircraft companies maneuvered to generate interest in their endless parade of proposals, the DoD, under direction from Congress and the Secretary of Defense, began to explore the options provided by the old fashioned process of prototyping. This approach, which effectively had been killed during the late 1950s when hardware costs began to soar seemingly out of control, now was taking on renewed importance because of the highly competitive elements of the acquisition process. This, coupled with the lengthier operational service careers that had become more commonplace for aircraft and other military vehicles, dictated that a more detailed and intense review of frontline hardware be given.

Northrop's enthusiasm for the idea of a fighter prototype fly-off program proved fairly guarded, as they were less than enthusiastic about the chances of their P-610 design. Regardless, the company found itself being forced slowly by circumstances beyond its control to significantly alter the P-530 from the design that had evolved over the preceding 5 years. By mid-1971, it was faced with an imminent AF prototype program calling for an austere 20,000 lbs.-or-less air superiority fighter with only a questionable chance for follow-on production.

Northrop also knew that, if such a program materialized, competition would be intense. In effect, Northrop had to be one of the winners or otherwise lose credibility with the large NATO market. Yet another concern was possible AF preference for a single-engine fighter. The single-engine P-610 was, of course, a hurried reaction to this concern, as was a program to pare no less than 5,000 lbs. off the twin-engine P-530's empty weight, but Northrop did not feel comfortable in submitting either design for competition.

After years of argument and political maneuvering both in the DoD and Congress, on August 25, 1971, LWF proponents were awarded their first major victory when a Program Decision Memorandum was unveiled approving hotly contested AF plans calling for the funding of two fly-offs: the Advanced Medium STOL Transport (AMST—later leading to the Boeing YC-14 and McDonnell Douglas YC-15 prototypes); and the Lightweight Fighter (LWF). Two competitive design and fabrication programs for "an operationally suitable, lightweight fighter" would be funded. Consequent to this, the Navy was directed "to monitor the (LWF) program to encourage carrier compatibility". This latter statement, though appearing almost inconsequential at the time, later would prove to have a long-lasting impact on the development of LWF hardware—particularly for the Navy.

During September, General Electric requested the AF fund a program to continue and accelerate the development of the YJ101 engine which, in fact, was very likely to be one of the competing engines in any forthcoming LWF competition. The proposed $10-million contract for this project would permit sufficient additional development to achieve a limited flight test rating within two years. The AF and GE had by now invested some $30-million in the YJ101 through research and development. Additionally, some 90% of the engine's detailed design had been released for manufacture.

The final decision concerning LWF prototype funding now rested with the U.S. Congress. Weeks of testimony covered virtually every facet of the proposed LWF effort and by the beginning of December, it had become apparent the tide finally had turned in favor of the program's feisty proponents. On December 14, 1971, the FY 1972 Appropriations Bill was released and $12-million was approved for "the initiation of prototype development of a lightweight fighter aircraft and a medium STOL transport aircraft." No other prototype programs were funded for any service.

In spite of the holiday season, the tasks remaining for program implementation were accomplished rapidly. On December 20, Assistant AF Sec. for Research and Development, Grant Hansen, approved the procurement actions, Dr. John Foster approved the Program Memorandum on the 27th, and Acting AF Sec. John McLucas authorized release of the Request for Proposals (RFP) and model contract to industry on the 31st. Thus on New Year's Eve, 1971, the AF launched the LWF program officially.

The LWF RFP was released to nine aerospace companies on January 6, 1972. It was requested that responses be returned no later than February 18. The amazingly short 21-page proposal contained performance and cost goals but few specifications. Limitations were placed on the length of the responses and the number of pages that could be devoted to cost, management, and aircraft description. Additionally, wind tunnel

A North American T-39D, BuNo. 150987, was modified by McDonnell Douglas and Hughes in order to provide an airborne testbed for the F/A-18's AN/APG-65 radar. This aircraft, operating from MCAIR's St. Louis facility, flew intercept missions against F-15s and F-4s, and also explored the unit's ground mapping and air-to-surface mission capabilities.

data was mandatory and a single wind tunnel model was required for evaluation.

Other general requirements included a high thrust-to-weight ratio, good transonic maneuverability, a simple avionics and fire control system package, a 6.5 g load factor, a suggested gross weight of no more than 20,000 lbs., and a suggested unit flyaway cost in 1972 dollars of no more than $3-million (based on a buy of 300 aircraft at a rate of 100 per year).

Five of the nine companies solicited, responded. Included were Boeing, General Dynamics, Lockheed, Northrop, and Vought. Their respective proposals were delivered to the AF on February 18, 1972, and a preliminary analysis was completed during the following three weeks. On March 18, the results were delivered to the Source Selection Authority (SSA) and shortly afterwards it was announced the Boeing Model 908-909 design was the number one contender, the General Dynamics Model 401 was a close second, and the Northrop P-600 (a twin-engine version of the P-610) was third (the single-engine P-610 originally had been submitted but had been withdrawn following an expressed foreign interest in the original P-530 by the Dutch).

The SSA analyzed the preliminary findings and after some revision, determined the General Dynamics proposal to be first choice, the Northrop proposal to be second, and the Boeing proposal to be third. The SSA based its conclusion on the fact the General Dynamics and Boeing proposals were quite similar in over-all appearance and thus, to satisfy one of the main objectives of the prototyping concept (the validation of emerging technologies), the decision was made to award the second contract to a different approach—represented by the Northrop design.

A final decision now was made by AF Sec. Robert Seamans, who declared the General Dynamics and Northrop proposals to be the two finalists. Accordingly, General Dynamics was awarded a contract for $37,943,000 for two Model 401s, and Northrop was awarded $39,878,715 for two Model 600s.

Following contract signing on April 13, 1972, the Northrop Corp. initiated construction of two YF-17 LWF prototypes. Gestation of these aircraft proved somewhat longer than that of the competing General Dynamics YF-16, due in part to their somewhat greater complexity and in part to delays in getting the GE YJ-101 engines approved for flight operations. Accordingly, the first prototype YF-17, 72-1569, wasn't rolled out of Northrop's Hawthorne, California facility until April 4, 1974.

April also saw the AF announce in favor of a hotly contested decision to make the LWF contract into something significantly more substantial than a technology demonstrator program. A production program had now been approved under the auspices of the Air Combat Fighter (ACF) title, this being the result of arguments in favor of a mix of both medium and high-technology fighters. Basically, it had been concluded that fighters such as the Grumman F-14 and McDonnell Douglas F-15 were so expensive they prohibited their respective military services from numerically meeting their respective force commitments. In order to accommodate these pre-conceived numbers, less expensive, and theoretically less sophisticated fighters, bought in large quantities, would permit force commitments to be met. This hi/low mix permitted a high level of technological sophistication while also allowing the quantity requirements to be achieved. The new ACF was the aircraft to fill this need.

The YF-17's design features, optimized to meet the demands for high energy levels, good maneuvering characteristics, excellent stall-post-stall characteristics, and turning capability in the air battle arena, included maneuvering flaps, large leading edge extensions (LEX), twin tails for directional stability, a cockpit designed for increased load factor tolerance, and a wing/engine inlet integrated to allow the engines to perform under the most difficult conditions of low airspeed, high altitude, and large yaw rates and high pitch angles.

The basic wing planform was tailored to provide good flying qualities and a high degree of spin resistance. The planform was combined with the large root leading edge extension (strake) which gave a remarkable increase in the maximum lift coefficient (it also reduced buffet intensity and decreased the drag encountered at supersonic speeds and during high maneuvering situations). The LEX approximately doubled the lift coefficient of the basic wing. The maneuvering flaps provided an additional increment of lift above the leading edge extension and basic wing planform.

The YF-17 also incorporated a rather unique feature known as differential area ruling. Normally, fuselage area

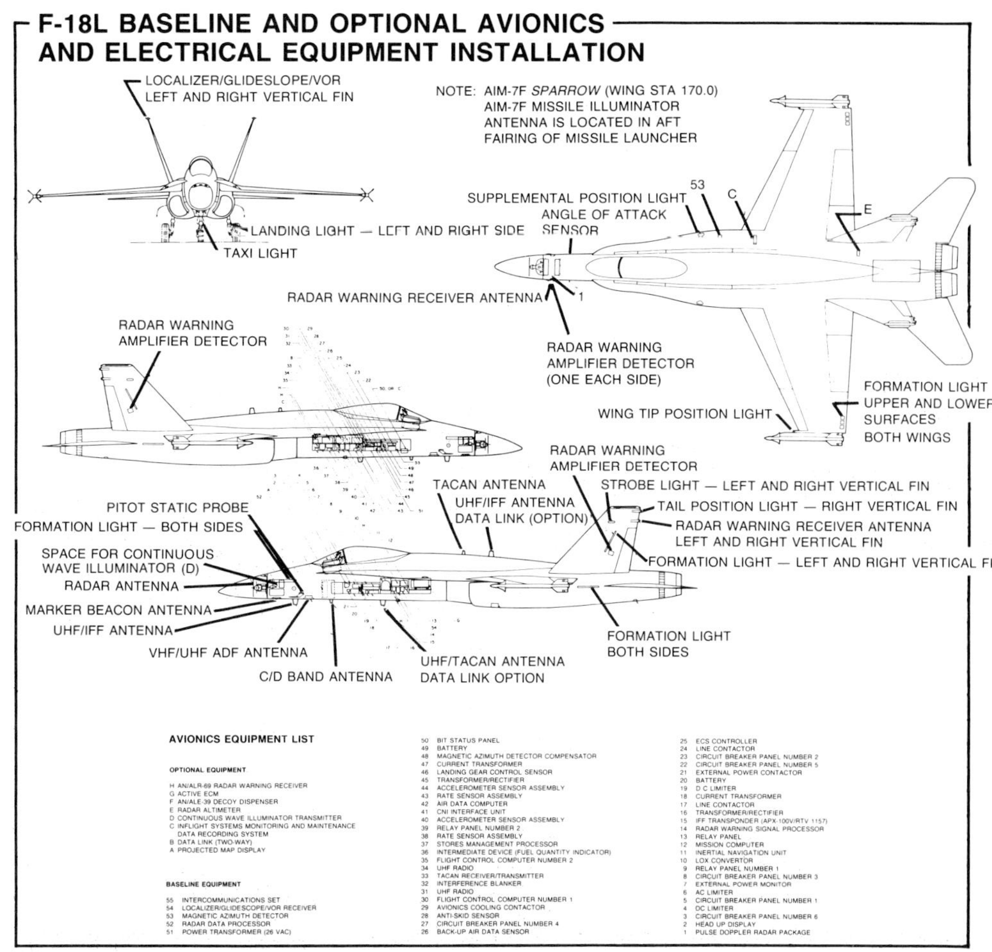

F-18L BASELINE AND OPTIONAL AVIONICS AND ELECTRICAL EQUIPMENT INSTALLATION

AVIONICS EQUIPMENT LIST

OPTIONAL EQUIPMENT

H	AN/ALR-69 RADAR WARNING RECEIVER
G	ACTIVE ECM
F	AN/ALE-39 DECOY DISPENSER
E	RADAR ALTIMETER
D	CONTINUOUS WAVE ILLUMINATOR TRANSMITTER
C	INFLIGHT SYSTEMS MONITORING AND MAINTENANCE DATA RECORDING SYSTEM
B	DATA LINK (TWO-WAY)
A	PROJECTED MAP DISPLAY

BASELINE EQUIPMENT

55	INTERCOMMUNICATIONS SET
54	LOCALIZER/GLIDESCOPE/VOR RECEIVER
53	MAGNETIC AZIMUTH DETECTOR
52	RADAR DATA PROCESSOR
51	POWER TRANSFORMER (26 VAC)
50	BIT STATUS PANEL
49	BATTERY
48	MAGNETIC AZIMUTH DETECTOR COMPENSATOR
47	CURRENT TRANSFORMER
46	LANDING GEAR CONTROL SENSOR
45	TRANSFORMER/RECTIFIER
44	ACCELEROMETER SENSOR ASSEMBLY
43	RATE SENSOR ASSEMBLY
42	AIR DATA COMPUTER
41	CNI INTERFACE UNIT
40	ACCELEROMETER SENSOR ASSEMBLY
39	RELAY PANEL NUMBER 2
38	RATE SENSOR ASSEMBLY
37	STORES MANAGEMENT PROCESSOR
36	INTERMEDIATE DEVICE (FUEL QUANTITY INDICATOR)
35	FLIGHT CONTROL COMPUTER NUMBER 2
34	UHF RADIO
33	TACAN RECEIVER/TRANSMITTER
32	INTERFERENCE BLANKER
31	UHF RADIO
30	FLIGHT CONTROL COMPUTER NUMBER 1
29	AVIONICS COOLING CONTACTOR
28	ANTI-SKID SENSOR
27	CIRCUIT BREAKER PANEL NUMBER 4
26	BACK-UP AIR DATA SENSOR
25	ECS CONTROLLER
24	LINE CONTACTOR
23	CIRCUIT BREAKER PANEL NUMBER 2
22	CIRCUIT BREAKER PANEL NUMBER 5
21	EXTERNAL POWER CONTACTOR
20	BATTERY
19	D C LIMITER
18	CURRENT TRANSFORMER
17	LINE CONTACTOR
16	TRANSFORMER/RECTIFIER
15	IFF TRANSPONDER (APX-100V/RTV 1157)
14	RADAR WARNING SIGNAL PROCESSOR
13	RELAY PANEL
12	MISSION COMPUTER
11	INERTIAL NAVIGATION UNIT
10	LOX CONVERTOR
9	RELAY PANEL NUMBER 1
8	CIRCUIT BREAKER PANEL NUMBER 3
7	EXTERNAL POWER MONITOR
6	AC LIMITER
5	CIRCUIT BREAKER PANEL NUMBER 1
4	DC LIMITER
3	CIRCUIT BREAKER PANEL NUMBER 6
2	HEAD UP DISPLAY
1	PULSE DOPPLER RADAR PACKAGE

ruling is performed to obtain higher transonic acceleration rates. However, for this aircraft it was felt that differential area ruling was desired for the upper and lower fuselage areas to optimize supersonic turning capability. The combination of all of these design features resulted in improved turn rates.

The twin vertical tails were canted outboard at an angle of approximately 20°. This feature provided the high degree of directional stability required during the dynamic maneuvers expected from this aircraft type. With the twin tails, one vertical always was expected to be in the free stream air to provide directional stability regardless of the angle-of-attack and sideslip magnitude.

Other YF-17 features included a Stencel Aero 3C ejection seat positioned in the cockpit at an angle of 18°; a bubble canopy providing excellent rearward visibility; a dorsally-mounted airbrake similar in intent and actuation to that found on the F-15; a fly-by-wire flight control system integrated with the ailerons, slab stabilators (which also were provided with mechanical pitch and roll back-up systems), rudders, and leading and trailing edge flaps; air-to-air capability in the form of missile pylons (nine external stores stations in total) and a General Electric M61 rotary cannon; a small Rockwell International ranging radar unit with its associated phased array antenna; a Litton Industries LN-33 inertial navigation system (INS); a Teledyne transponder for navigation; and a JLM International head-up display (HUD).

The YF-17 was powered by two General Electric YJ101 two-shaft afterburning turbojet engines rated at approximately 15,000 lbs. th. each. Born as the GE15, this engine eventually would enter production as the F404.

A considerable amount of preflight phase testing was accomplished in the YF-17 program prior to the actual inflight program initiation. The facilities utilized to accomplish this testing were the large amplitude flight simulator located at Northrop's Hawthorne, California production plant, the spin and free flight test tunnels at Langley, and the CALSPAN (New York) variable stability T-33 aircraft.

Wind tunnel development activities related to the YF-17 extended over a period of eight years and amounted to 9,700 hours total time. Significant features developed in the wind tunnel were the hybrid wing and the engine/airframe integration, as well as the positioning of the vertical and horizontal tail surfaces for optimal high angle-of-

The basic P-530 configuration, predecessor to the YF-17 and later, F/A-18, went through a number of full-scale mock-up studies that included (left) an early configuration with abbreviate LEXs and half-cone intake spikes. The final "Cobra" design study (right) was, to all intents and purposes, the full-scale F/A-18 mock-up.

The first (right) and second F/A-18A prototypes, BuNos. 160775 and 160776, respectively, during the course of preliminary flight test work at NATC Patuxent River, Maryland. The second aircraft lacked the gold trim and enhanced markings of the first.

The third F/A-18A FSD aircraft, BuNo. 160777, was utilized for carrier suitability trials. It is seen during 1979 land-based catapult and trap tests at Patuxent River. Blue and white markings were accented in select places by thin gold cheat lines.

The fourth F/A-18A FSD aircraft, BuNo. 160778, was utilized for inflight structural and loads testing and accordingly was utilized to explore airframe deformation characteristics while carrying a wide variety of external loads.

The fifth F/A-18A FSD aircraft, BuNo. 160779, was utilized for avionics and weapons system testing and was the first "Hornet" to be equipped with the Hughes AN/APG-65 radar. Unpainted dielectric nose radome is noteworthy.

attack performance.

The stall-post-stall wind tunnel results indicated a stable, controllable aircraft at angles-of-attack as high as 45° and recovery from spins that could be accomplished with normal controls—even from the flat spin mode.

The large amplitude flight simulator was utilized to investigate flying qualities in the normal and failure modes of the flight control system, to develop tracking capabilities and criteria, and to explore stall-post-stall aircraft characteristics. During the accomplishment of these investigations, the USAF LWF test team pilots from the Air Force Flight Test Center (AFFTC) and the Tactical Air Command (TAC) participated in all of the evaluation activities conducted. Additionally, all test force pilots including the contractor pilots flew the CALSPAN variable-stability Lockheed T-33 which allowed a real world assessment of approach conditions and selected-point-in-the-sky type evaluations.

The initial airborne testing of the YF-17 began with the first flight of aircraft 72-1569 at Edwards AFB on June 9, 1974, with Northrop company test pilot Hank Chouteau at the controls. The flight lasted a total of 61 minutes and also served as the first flight of the GE YJ101 engine. As the program now rapidly progressed from one of pure evaluation to one of competitive aircraft selection, the flight test program was prioritized to define the areas where inflight data was required to support a decision related to aircraft choice.

On August 21, 1974, the second YF-17, 72-1570, also

with Northrop Company test pilot Hank Chouteau at the controls completed its first flight from the Edwards AFB main runway. By this time, the competition between the General Dynamics and Northrop aircraft was well underway, with General Dynamics having the distinct advantage of having had both its prototypes airborne throughout the competition schedule. This allowed the YF-16s to generate considerably more sortie time than their Northrop competitors (which were hindered doubly by the fact that their YJ101s were unproven engines).

The competition proved short and concise. By the time the fly-off was ended, a relatively large number of pilots had been permitted stick time in both aircraft types. Company and AF test pilots had logged the majority of hours, with Navy and miscellaneous government pilots claiming the rest. Inputs from these professionals played a key role in determining the competition winner.

Military pilots assigned to the LWF fly-off were given an opportunity to fly both the YF-16 and YF-17, and usually more than once. Additionally, following the numerous air-to-air combat sessions with a variety of adversary aircraft types (including the Cessna A-37B, the McDonnell F-4E, the Convair F-106A, the MiG-17, the MiG-21, and the MiG-23), adversary pilots were asked to critique the two designs in terms of competitiveness.

During the course of flight testing, the YF-17 was found to perform satisfactorily and per Northrop spec. It demonstrated a top speed of Mach 1.95, a peak load factor of 9.4 g, a maximum altitude of just over 50,000 ft.,

and a sea level rate of climb exceeding 50,000 ft. per minute. Pilots found the aircraft pleasant to fly and with few serious control vices. It could achieve a controllable angle-of-attack of 34° in level flight and a 63° angle-of-attack during a 60° zoom climb. Control could be maintained at minimum indicated airspeeds of as little as 20 kts. Such performance later led Northrop to claim the YF-17 had no angle-of-attack limitations, no control limitations, and no adverse departure tendencies.

The last missions of the LWF competition took place during mid-December 1974, and the following several weeks were spent completing analysis of gathered data and pilot input. Finally, on January 13, 1975, Secretary of the AF John McLucas announced the General Dynamics YF-16 as the LWF (Air Combat Fighter—as it now was being called) program winner based on the program's two primary criteria, performance and cost. Importantly, the YF-16 had been the almost unanimous choice of competition pilots.

The loss proved a significant set-back for Northrop, though one not totally unexpected. The company had been somewhat dissatisfied from the first with the way the fly-off had been conducted, and in fact had raised serious doubts about several important judging criteria. Some analysts later noted that competition results almost certainly would have favored Northrop if the LWF fly-off had remained a paper exercise rather than hardware.

The analysts based their claims on the following:

(1) Northrop had proposed a bigger aircraft and it is likely that the "bigger aircraft have more capability" syndrome, then so commonplace in the DoD, would have prevailed.

(2) Northrop's aircraft, though appearing more expensive, was claimed actually to cost less. And Northrop's reputation for building inexpensive aircraft likely would have influenced the decision-making process.

(3) The Northrop design was technologically less sophisticated and therefore less of a risk. Conservatism consistently has prevailed in competitions such as that for the LWF.

Northrop also felt it had operated at a distinct disadvantage due to the fact the YF-17 had entered the flight test program four months later than the YF-16, thus forcing the company to cram all testing into six months rather than ten (NATO was awaiting the final LWF competition verdict before initiating the final stages of its decision making process—and it had requested the AF make a choice no later than December 1974—thus forcing the competition deadline); and the YF-17 was flying with a prototype engine (the YJ101) whereas the YF-16's F100

The sixth F/A-18A FSD aircraft, BuNo. 160780, was the first to be painted in red-on-white markings. This aircraft was utilized for spin and high angle-of-attack tests and thus was given markings optimized for high visibility. Following delivery to Patuxent River, it was equipped with an explosively deployed anti-spin chute.

The seventh F/A-18A FSD aircraft, BuNo. 160782, was utilized for armament and systems test work and accordingly, was equipped with a functioning Hughes AN/APG-65 radar and miscellaneous weapons related monitoring systems.

The eighth F/A-18A FSD aircraft, BuNo. 160783, was utilized as a performance and systems testbed. It is seen during gun gas ingestion trials of its nose-mounted M61A1 rotary cannon. Gun smoke pattern proved nominal for both pilot and engines.

The first prototype TF-18A (later F/A-18B), BuNo. 160781, during a test hop out of the McDonnell Douglas St. Louis facility and prior to its delivery to Patuxent River. This aircraft was painted virtually identically to the first F/A-18A (BuNo. 160775).

The first production F/A-18B, BuNo. 160784, taking off at the beginning of its 1980 Farnborough Airshow routine. This aircraft was destroyed on September 8, the day after, when the right engine failed shortly after departure for Spain.

was a proven powerplant with significant operational time.

Actual factors effecting the final LWF decision included the noteworthy differences in estimated production costs (the YF-17's flyaway costs were determined by AF analysts to be some $250,000 more per aircraft than those for the YF-16) and operating costs; the YF-16's distinctly superior transonic and supersonic performance; the YF-16's decidedly superior maneuverabilty envelope; the YF-16's greater excess thrust; and the YF-16's superior range—which, in fact, later proved to be comparable to that of the McDonnell Douglas F-15.

The following table lists the physical and performance characteristics of the YF-16 and YF-17 at the time of the LWF competition:

	YF-16	YF-17
Wingspan	30'0''	35'0''
Length	46'6''	56'0''
Height	16'3''	14'6''
Wing sweep @ leading edge/ quarter chord	40° (LE)	20° (QC)
Aspect ratio	3:1	3.5:1
Wing area	280 sq.'	320 sq.'
Empty weight	14,023 lbs.	17,560 lbs.
Gross weight	20,665 lbs.	26,960 lbs.
Max. gross weight	31,000 lbs.	38,000 lbs.
Fuel weight	6,642 lbs.	9,400 lbs.
External load	13,200 lbs.	18,000 lbs.
Combat thrust-to-weight w/afterburner	1.28:1	1.2:1
Combat wing loading	75 lbs./sq.'	78 lbs./sq.'
Powerplant	1 x P&W F100-PW-100	2 x GE YJ101-GE-100
Thrust rating	23,500 lbs.	14,750 lbs. each
Bypass ratio	0.7	0.20
Combat radius	600 mi.	576 mi.
Ferry range	3,000 + mi.	2,780 mi.

The F-16 program now moved ahead rapidly under AF auspices while Northrop and the YF-17 were left to their own devices. Accordingly, Northrop continued pursuing with some urgency its remaining options, these including a long-shot contract with the Navy, and similarly long-shot production programs for non-indigenous air forces. By early 1975, the YF-17's future looked decidedly bleak.

THE NAVY:

During the early days of the LWF program, it had become apparent to proponents of the basic LWF philosophy that there was an obvious demand for a similar aircraft to fulfill certain Navy requirements. There was, in fact, a Navy LWF team operating in the Naval Air Systems Command, and the successes enjoyed by

the AF's LWF proponents gave them strong incentive to continue their push for a Navy aircraft of similar design.

Like the AF LWF, the birth of the Navy LWF had been brought about not only by need, but also by a new Navy fighter competition (VFX) calling for yet another expensive, large, and extraordinarily complex fighter. This program, the result of a decision to avoid acquisition of the problem-plagued bi-service General Dynamics F-111 (the Navy's version was the Grumman-manufactured F-111B), was soon to become the F-14—and as such, the most expensive and complex production fighter aircraft in history.

With the birth of the VFX program, certain factions within the Navy began to awaken to the fact the service had two serious problems: (1) its new fighter, at what eventually would amount to over $30-million per copy, was quickly and noisily eating up the Navy acquisition budget; and (2) there was a strong move in Congress calling for an increased force structure (i.e., more combat vehicles were needed). In response to the latter, various House and Senate members had begun openly to advocate the aforementioned hi/low mix, primarily out of fear the Soviet Union's military arsenal was growing more rapidly than U.S. technology advantages.

For the Navy, this interest in increased numbers spelled only one thing—sooner or later the service would be forced to look beyond the expensive F-14 and be forced to accept a fighter that was less complex and cheaper (and, in the Navy's opinion, probably less capable).

During the summer of 1973, James Schlesinger took over as the new Secretary of Defense. As such, he was not long in becoming embroiled in the controversy surrounding the LWF philosophy. Through good fortune, he also was not long in being introduced to the more staunch supporters of the AF LWF program, including Boyd, Sprey, Meyer, and Riccioni. These latter quickly indoctrinated him and in short order, made him a strong LWF supporter.

Schlesinger now mandated that both services (AF and Navy) study LWF design with significant intensity. During August 1973, the Navy was instructed by Congress to pursue a study calling for a lower cost alternative to the F-14, and the following month, they were ordered to request proposals from industry. Several Navy study groups had anticipated the mandate and were well underway with studies of their own by the time the official dictum was released. Unfortunately, like many of their

F/A-18B, BuNo. 161704, of VMFAT-101. This unit, which was created to accommodate expanding "Hornet" training requirements, is one of the most recent Marine Corps squadrons to form. SH tailcode was selected for descriptive call sign options.

F/A-18A of VMFA-115. Subdued markings with eagle logo on tail consist of gray on light compass ghost gray. Tail code is visible as small "VE" on tip of vertical fin. Miscellaneous markings, including fuselage insigne, consist of grays reversed out.

Contemporary VMFA-115 markings offer greater contrast between the grays and thus are slightly more accentuated than those seen on earlier squadron aircraft. Still newer markings include an eagle of significantly smaller dimensions.

VMFA-122 F/A-18A, BuNo. 161952, at MCAS Beaufort during September 1986. Aircraft bears standard insigne, VMFA-122 tail code, and badge in dark gray. Two-tone gray camouflage is standard, as are all-white landing gear strut assemblies.

predecessors, these inbred study teams were still of the opinion the F-14 alternative would have to be equipped with a Beyond Visual Range (BVR) radar and associated weapon system. This meant the aircraft was to be equipped with radar-guided missiles.

The LWF proponent group had opposed the use of this armament type from the beginning. Their rationale was that radar guided missiles and their associated systems were expensive; the size of the required radar would affect aircraft performance (BVR systems require a large and heavy radar dish—which consequently increases the aircraft's frontal area; in order to offset these two factors, larger and more powerful engines are required to propel the aircraft, which in turn imply a greater fuel load requirement and a further increase in weight); Identification Friend or Foe (IFF) systems were relatively ineffective and not to be trusted in actual combat: BVR sytems had an extraordinarily poor success record in actual combat; high-g maneuvering during air combat tended to break BVR radar-guided missile lock-ons; and radar-guided missiles take critical time to stabilize during a lock-on and thus commit the carrier aircraft to windows of vulnerability far in excess of the acceptable time constraints dictated by air combat.

The Navy, in an attempt to protect the F-14 from funding incursions, maneuvered its way through a number of additional studies and finally presented a proposal to the Congress and Schlesinger calling for an austere, "stripped" version of the F-14 to be called the F-14X. This aircraft deleted the AIM-54 *Phoenix* air-to-air missile and its associated weapon system (around which the aircraft originally had been designed!), but retained the AIM-7 *Sparrow* radar-guided AAM system. It was, so the Navy claimed, the best and cheapest solution to the increased force structure problem.

Congress and Schlesinger were quick to disagree. On May 10, 1974, the House Armed Services Committee deleted the Navy's request for $34-million to start their new VFAX program which called for a totally new fighter to complement the F-14, and instead suggested the Navy consider adopting one of the two AF lightweight fighters (YF-16 or YF-17) then undergoing preliminary flight testing at Edwards AFB.

The Navy, of course, reacted adversely, but following additional ill-fated attempts to push the VFAX program through, quietly gave-in to Congress and Schlesinger when the powerful Senate Joint Committee on Appropriations released a strong statement in support of the Navy LWF buy.

Thus it was that the Navy found itself in the position of having to buy whichever LWF design the AF found most to its liking. It was a bitter pill for the Navy to swallow and, following the traumatic F-111B experience of the early 1960s, one that was not likely to go down without serious argument.

While the Navy explored its few remaining options and debated the Congressional dictum internally, Northrop reached a partnership agreement with the McDonnell Douglas corporation concerning a navalized version of Northrop's YF-17. Similarly, General Dynamics and Vought joined in an agreement that would give Vought the right to produce a navalized version of the YF-16.

Not surprisingly, the Navy, almost from the very beginning, had taken a strong stand in favor of the YF-17. The service particularly liked the Northrop aircraft's two engines, its apparent room for airframe growth and expansion, its apparent mission adaptability, and the fact that McDonnell Douglas, a successful producer of Navy fighters, had been picked by Northrop as its Navy partner. During the course of the ACF fly-off, Navy pilots flew both the YF-16 and YF-17 prototypes and concluded the YF-17, though modestly deficient in performance, was the preferable design for Navy needs.

When the final LWF/ACF competition winner, General Dynamics and their YF-16, was announced by the Secretary of the AF on January 13, 1975, the Navy was livid. The response was universally negative. Informally, the Navy stated it had no intention of following the mandate set forth by Congress and Schlesinger, and it immediately embarked on a campaign either to kill the Navy LWF program or to force Congress to allow the service to acquire the navalized YF-17.

During the four months following the AF award to General Dynamics, the Navy maneuvered with Congress and the Secretary of Defense and eventually reached a compromise agreement, on May 2, 1975, calling for the service to develop a derivative of the Northrop YF-17 (known originally as the P-630 and later as the McDonnell Douglas Model 267), designated originally as the F-18 Navy Air Combat Fighter (NACF) and later as the F/A-18 Navy/Marine Corps "strike fighter" (the latter denoting its dual-mission capability). The initial plan provided for distinctive F-18 fighter and A-18 attack versions, but later all hardware differences were eliminated (thus causing the fighter and attack variants to differ only in the armaments carried for specific missions; originally, the attack version of the F/A-18 was to have been designated F-18L; Northrop would have been the primary contractor, with MCAIR being the primary subcontractor—with the split being 60%/40%, respectively).

While the Navy's decision incurred considerable congressional criticism during 1975, there was gradual ac-

ceptance of the Navy's contention that technical changes required to make the YF-16 carrier-suitable would negate the commonality between AF and Navy fighters sought by Congress and could be less cost-effective than development of a different aircraft based on the YF-17 design. During 1976, full-scale development contracts were awarded by the Navy to McDonnell Douglas for airframe construction and final assembly (with Northrop as the major subcontractor, responsible for approximately 40% of the airframe) and to General Electric for the F404 engine.

Northrop's decision to step down as the prime contractor in the program was the product of many internal and external forces, all of which were spearheaded by the company's lack of experience in building carrier-suitable hardware. Additionally, major programs more in line with Northrop's production abilities were looming on the horizon, and the company did not have the physical plant to handle the production capacities that eventually would be demanded for several programs. Though lawsuits resulting from McDonnell Douglas' attempts to sell their version of the *Hornet* to foreign customers were initiated by Northrop, these were reconciled during the course of litigation lasting some six years. A conciliatory agreement signed on April 8, 1985 allowed the two companies to successfully accommodate production obligations.

Procurement rates were to change significantly following the Navy's decision to produce the new aircraft, these increasing from 5 to 9 during FY 79, from 15 to 25 during FY 80, and from 48 to 60 during FY 81. By the middle of 1981, the Navy had made the F/A-18 program into its largest aircraft acquisition, with a commitment to buy no less than 1,377 (including 11 RDT&E and 1,366 production samples). As of the date of this writing, just over 1,500 orders for the F/A-18, including foreign sales, stand extant.

Once the Navy had committed itself to the McDonnell Douglas/Northrop aircraft (with production responsibilities effectively split 60/40 between the two, respectively Northrop would build the vertical tail assemblies and the center and aft fuselage subassemblies, and MCAIR would build all remaining components including the wings, while serving also as the final assembler) a major redesign effort was undertaken to accommodate the Navy's (and Marine's) specific operational requirements MCAIR received a letter contract confirming contract consummation on January 22, 1976, this calling for eleven Full-Scale Development (FSD) aircraft (9 single-seat and 2 two-seat) and a tentative first flight date of July 1978

Consequent to this, and actually preceding it by several

Line-up of VMFA-122 F/A-18As. These aircraft now are stationed at MCAS Iwakuni with MAG-15. Two-tone gray camouflage is standard for type. All miscellaneous markings, including tail code and insigne, are dark gray.

F/A-18A, BuNo. 161971 of VMFA-122 at MCAS Beaufort on September 16, 1986. FOD screens are barely discernible attached to the engine intakes. A horizontal ejector rack is attached to the single wing pylon.

F/A-18A, BuNo. 161961, of VMFA-251 on September 16, 1986 at MCAS Beaufort. Subdued markings are barely visible in the form of a lightning bolt and DW code on the vertical fin and rudder. Extended cockpit access ladder is noteworthy.

F/A-18A of VMFA-251. The "Thunderbolts" transitioned from McDonnell F-4J/Ss and stood up at MCAS Beaufort during June 1986. Aircraft is seen undergoing preflight inspection by ground personnel.

months, was the preordained decision to utilize General Electric's advanced YJ101 turbojet engine, designated F404, in the aircraft. This engine, which represented the state-of-the-art in conventional turbojet engine design, was significantly more powerful than its predecessor, and fully capable of providing the propulsion needed by the new Navy fighter.

At the time of contract consummation, the new, Navalized YF-17 was redesignated F-18 (later, F/A-18 once the two roles/two versions program was combined). This was done not necessarily as a matter of course, but rather as an attempt to represent the aircraft as a totally new design—unrelated to its AF forebear. In fact, it essentially was new, with few components related to those found on the YF-17. Major changes, which would have a decidedly serious effect on the external configuration of the aircraft as finally built, included the addition of a Hughes-manufactured multi-mode radar; provision for the transport of the large, radar-guided Raytheon-manufactured AIM-7 *Sparrow* air-to-air missile and its associated fire control system; provision for 4,460 lbs. of additional fuel to accommodate long range patrol requirements (giving the aircraft a total of 10,680 lbs. of fuel in four fuselage tanks and two inboard wing section tanks); an articulated nosewheel towbar; an arrester hook; folding outer wing panels; increased wing area (2.5 ft. was added to the YF-17's wingspan, and the chord was increased to accommodate the necessary aspect ratios; total wing area thus was increased by 50 ft. over that of the YF-17's 350 sq. ft.) to accommodate the increased wing loading brought on by additional weight; general structural strengthening all around to accommodate the rigors of carrier operations; and a completely redesigned and significantly strengthened landing gear. These changes added no less than 10,580 lbs. to the YF-17's 23,000 lb. gross weight.

Other changes were optimized to improve the aircraft's performance in the low speed and maneuverability portions of the flight envelope. Included were refinements in the unorthodox wing leading edge extensions (involving redesign and increases in total area); the deployment angles of the leading and trailing edge flaps which were increased from 30° to 45°; the ailerons which were programmed to droop at a maximum angle of 45° in low speed flight; the stabilators which were revised in planform (resulting in more area, but a lower aspect ratio); and a dog tooth which was added to the leading edges of both the wings and the stabilators to cause vortex generation (this curtailing a spanwise-flow anomaly that was affecting adversely the aircraft low-speed controllability). Approach speeds now were predicted to be approximately 125 kts. at an angle-of-attack of 6° to 7°, thus improving pilot visibility and wind-over-deck performance. Consequent to these improvements, the F/A-18 was designed to have an active service life of 6,000 flight hours, including 2,000 catapult launches and 2,000 traps.

It is interesting, at this point to compare the Navy's original VFAX requirement to those projected for the then-forthcoming F/A-18:

	VFAX	F/A-18
Vmax (non-afterburning)	.98/1.0 Mach	.99 Mach
Acceleration from Mach .98 to 1.6	80 to 100 secs.	88.3 secs.
Combat ceiling	45,000 to 50,000 ft.	49,300 ft.
Ps at Mach .9/10,000 ft.	750/850 ft. per sec.	756 ft. per sec.
Buffet-free sustained load factor	5 to 5.5 g	6.6 g
Structural load factor	7.5 g	7.5 g
Single engine rate of climb	500 ft.	565 ft.
Min. approach speed	115 to 125 kts.	131 kts.
Escort radius (w/conventional payload)	400 to 450 mi.	415 mi.
Strike radius (w/conventional payload)	550 mi.	655 mi.

By mid-1978, construction of the first F/A-18 prototype, BuNo. 160775, nearly was complete, and on September 13, 1978, it was rolled out formally from the final assembly building at MCAIR's Lambert Field, St. Louis facility for the first time. Bearing Navy markings on its left side and Marine Corps markings on its right, and with an overall white, blue, and gold paint scheme, this aircraft completed the first F/A-18 flight just over two months later, on November 18. MCAIR test pilot Jack Krings was at the controls as the aircraft departed St. Louis just after 1100 hours local time. Airborne for some 50 minutes, it logged a maximum speed of 300 kts. and an altitude of 24,000 ft.

Additional test hops now followed at St. Louis in rapid succession, with the prototype then being moved to the Naval Air Test Center at Patuxent River, Maryland during early January 1979. The NATC assigned the aircraft to basic flight test and flutter exploration flights as part of the *Hornet* FSD program which was scheduled to be ongoing from January 1979 to October 1982. Eventually, ten other F/A-18s would be assigned to the FSD program, these including the following:

BuNo.	A/C –	Test Function
160776 (YF/A-18A)	2	Propulsion and performance
160777 (YF/A-18A)	3	Carrier suitability and ECS
160778 (YF/A-18A)	4	Structural flight test
160779 (YF/A-18A)	5	Avionics and weapons systems
160780 (YF/A-18A)	6	High AoA and spinning
160781 (YF/A-18B)	T1	Armament and systems
160782 (YF/A-18A)	7	Armament and systems
160783 (YF/A-18A)	8	Performance and systems
160784 (YF/A-18B)	T2	Accelerated engine service test
160785 (YF/A-18A)	9	Maintenance engineering

Under the new Principal Site Concept philosophy of flight testing, almost all F/A-18 flight test work was conducted at Patuxent River. This facilitated Navy input into the flight test and development program and allowed MCAIR to incorporate changes and modifications in a more timely and economic fashion.

FSD flight testing initially called for no less than 3,257 flights during a two shift, 6 days a week schedule. By mid-1981, more than 3,500 hours had been logged during more than 2,600 flights. Amid significant political controversy, the flight tests were conducted at a steady pace, laboriously exploring the aircraft's flight envelope and consequently exposing numerous technological and performance failings.

Emerging problems with the aircraft and their forced public acknowledgement via the media helped fuel rumors of program cancellation while providing fodder for pundits' cannons. Additionally, flight test reports leaked to the press underscored pilot concerns about the F/A-18's performance and its ability to accomplish mission objectives. Among the shortfalls eventually brought under public scrutiny were:

(1) Unacceptably high nose wheel lift-off speeds—the F/A-18 prototypes were found to require 140 kts. for rotation. The solution was to modify the stabilator leading edges by removing a MCAIR-designed dog tooth, and revise the control system software. The latter provided an input which automatically generated a 25° rudder toe-in during takeoff while weight still was concentrated on the main landing gear. The toe-in provided an additional downward moment aft of the aircraft rotational axis and thus reduced the nosewheel lift-off speed to a more palatable 115 kts.

(2) Unacceptable performance deficiencies—the F/A-18 prototypes, during the Navy Preliminary Evaluation program, were found to have 12% less range than guaranteed by MCAIR. A similarly large deficiency was found in acceleration from Mach .8 to Mach 1.6 at 35,000 ft. (110 seconds had been guaranteed). Additionally, roll rates, initially spec'd at 280° per second, were substandard, maneuverability failings were significant, and the environmental control system was inefficient.

These latter failings required significant study and effort on MCAIR's behalf, and it wasn't until numerous changes and modifications had been incorporated that most were overcome. Solutions included propulsion system upgrades, leading and trailing edge flap computer programming upgrades (the fllaps were found to be 2°

F/A-18A, BuNo. 161978 of VMFA-251 at MCAS Beaufort. Variation in vertical fin markings is readily apparent in this view, with aircraft in background bearing lightning strike and DW code in lighter, possibly more weathered, paint.

F/A-18A, BuNo. 161967, of VMFA-251 at MCAS Beaufort on September 16, 1987. "Remove Before Flight" warning flags can easily be discerned hanging from the landing gear wells and exhaust nozzle areas.

F/A-18A, BuNo. 163166, of VMFA-312, on final approach to NAS Dallas on May 8, 1988. Vertical fin has a gray on dark gray checkerboard with a top border stripe in yellow and a bottom border stripe in red. All other markings are standard for type.

F/A-18A, BuNo. 161757, of VMFA-314, transient at Williams AFB on December 9, 1984. Particularly noteworthy in this view is the in-use single-point ground refueling receptacle. All tanks in the aircraft can be filled via this connection.

to 3° leading-edge-down off in cruise—thus increasing cruise drag), a significant ECS upgrade, and stiffening of various structural components including the outer wing panels and their associated attachment assemblies.

In an attempt to reduce aircraft drag, tuft studies were conducted to visually determine the effects of filling-in the slots in the wing leading edge extensions, and this modification later became a standard item on production aircraft. Additionally, the wing leading edge radius was increased and a fairing was installed over the ECS exhaust port forcing the efflux rearward.

The roll rate problem initially proved perhaps the most serious failing of all. Achieved roll rates during flight test included 185° per second at Mach .7, 160° per second at Mach .8, and 100° per second at Mach .9, all at 10,000 ft. At 20,000 ft. the roll rate was per spec. at Mach .9, but decreased as the aircraft accelerated into the high transonic and low supersonic zone.

The cause of the roll performance deficiency quickly was traced to two failings: flexing of the outer wing panels in high load situations, and too much roll damping when the wingtip-mounted *Sidewinder* air-to-air missiles were in place.

The flex anomaly was found to result in a form of aeroelastic divergence (i.e., when the aileron moved either up or down, the wing leading edge flexed in the opposite direction due to wing twist, thus partially negating the aileron's aerodynamic rolling moment effect). To counter this, MCAIR embarked immediately on a modification program that eliminated the outer panel's dog-tooth leading edge and led to the development of a strengthened structural assembly that increased torsional rigidity. Monolithic graphite composites now replaced the inner wing sandwich material and the aluminum outer panel material. Additionally, the wing trailing edge spar assembly was thickened and thus strengthened, and the associated web and cap assemblies also were modified. In order to offset the weight increase and consequently improve the roll rate of the aircraft, the ailerons were extended outboard to the wingtips, thus providing 36% more area (and consequently leading to a desirable 7 kt. reduction in landing speed), and differential movement of the leading and trailing edge flaps was incorporated. Finally, the horizontal stabilator differential authority (i.e., roll rate) also was increased.

These changes improved the maximum roll rate to 220° per second and thus brought roll performance up to a competitive level. To date, variations in this near-spec capability throughout various parts of the envelope remain contentious, but are not considered a major deficiency.

The range deficiency problem proved significantly more difficult to overcome. The original Navy spec. called for a range of 444 n. miles with the aircraft configured as a fighter, and 635 n. miles with the aircraft configured for ground attack. Tests during November 1979, utilizing various FSD aircraft, generated actual figures of 404 n. miles and 580 n. miles, respectively. Though MCAIR quickly suggested that additional external and internal fuel tanks would allow the aircraft to meet the spec., the Navy declined after noting that the spec'd weight limit of 20,146 lbs. already had been exceeded by no less than 1,962 lbs.

These improvements did not fully correct the performance anomalies discovered during the FSD flight test program, but they did narrow the gap. To date, range remains approximately 8% short of spec., specific fuel consumption shortfalls remain throughout the envelope, and acceleration performance continues to be less than desired.

While efforts to solve the F/A-18's performance shortfalls continued at MCAIR, the Navy, during June 1979, took BuNo. 160777 and initiated carrier compatibility trials at Patuxent River. Immediately prior to the initiation of actual trial activity, MCAIR had concentrated testing on proving the F/A-18 structurally sound for carrier catapults/arrestments and making the aircraft pilot compatible. Concurrently, the operating restrictions of the F/A-18 were expanded to allow a reasonable initial sea trials effort. Specifically, the maximum sink rate was expanded to 18 + fps, and catapult capability was expanded from no capability to a maximum catapult longitudinal acceleration of 5 g's. Additionally, the aircraft was given a limited instrument flight rules capability after water ingestion and icing tests behind a KC-135A water spray tanker, and electromagnetic compatibility testing was conducted at the Naval Air Test Center successfully demonstrating that the F/A-18 could survive the electromagnetic environment of a carrier flight deck. Finally, thermal and sonic environment tests were conducted at the Naval Air Engineering Center, Lakehurst, New Jersey to ensure the aircraft could function in its own noise and vibrations in front of a carrier's jet blast deflector.

At Patuxent River, over a period of several weeks, 70 land catapult launches and 120 arrested landings were completed before the aircraft was flown by Navy test pilots Lt. Cdr. Dick Richards and Lt. Ken Grubbs to the U.S.S. *America* for initial sea trials beginning on October 30, 1979. The following four days led to 32 catapult launches and traps, and 17 touch and go landings. Vertical descent rates of up to 19.5 fps were demonstrated

F/A-18A of VMFA-314 at the moment of launch from the port catapult of the USS "Constellation" during July 1983. Nose gear catapult link bar can be seen attached to shuttle assembly. Noteworthy is toed-in rudder position which is automatically pre-set as part of the control system take-off configuration.

F/A-18A of VMFA-323 on final approach to NAS Dallas during 1987. Vertical fin band is black with gray diamonds. Tail code and squadron identifier are also in black. All other markings are standard for type.

F/A-18A of VMFA-323 shortly after takeoff. Visible attached to the starboard intake tunnel station is a Martin Marietta laser spot tracker/strike camera (LST/SCAM) pod and attached to the wing pylon is a single laser-guided 500 lb. bomb.

Transient F/A-18A, BuNo. 163155 of VMFA-333 taxiing out for takeoff from Ellington AFB on January 31, 1988. Tail code and shamrock markings are all dark gray on gray. Toed-in position of rudder, even during taxi, is noteworthy.

F/A-18A, BuNo. 163133, of VMFA-451, on April 21, 1987. When hydraulics are unpowered, the ''Hornet's'' flaps, flaperons, and miscellaneous other control surfaces tend to stabilize in a drooped position.

successfully.

By the time the carrier trials were concluded, the F/A-18's suitability was well documented and few criticisms were voiced. Most of the catapult launches had been made with the engines at military power, two were made with the engines in afterburner, and no major difficulties surfaced when launching the aircraft using either setting. A nose gear failure leading to the redesign of a shrink link lug proved the only major aberration of the tests.

Following the trials and upon returning to Patuxent River, BuNo. 160777 suffered a main landing gear failure during touchdown. The fault was traced to the main gear axle centering unit and a fatigue anomaly stemming from stress dynamics generated at touchdown. A dual-chamber shock strut solved the problem and was retrofitted to extant F/A-18s and immediately incorporated into the production line.

Following completion of repairs during the early summer of 1980, BuNo. 160777 was flown to Edwards AFB where crosswind landing trials were undertaken. A total of 119 landings were completed successfully in crosswinds of up to 30 mph. Additional carrier compatibility related trials included the first fully automatic landing which was conducted at Patuxent River on January 22, 1982, with MCAIR test pilot Peter Pilcher in the cockpit. Pilcher did not touch the aircraft's controls during the final approach and touchdown sequence, thus completing the F/A-18's first real test of its auto-land option. Four days after the initial trial, other Navy pilots were introduced to the system, and by August the confidence level was high enough to move further testing out to sea. Accordingly, the U.S.S. Carl Vinson was given the honor of hosting the second series of F/A-18 carrier suitability trials and over a period of several weeks, some 63 catapult launches and traps (with many bolters) were completed successfully during both day and night while the aircraft was carrying a wide variety of external stores combinations.

Weapons system trials exploring the air-to-air and air-to-surface capabilities of the aircraft were conducted with several of the other FSD airframes, including BuNo. 160779, 160781, and 160782. The first live missile firing, using an AIM-9, was conducted by BuNo. 160779 during December 1979, with MCAIR test pilot Bill Lowe having the honor. The missile scored a near miss and effectively would have destroyed its BQM-34 target if its warhead had been functional. Eight additional firings took place over the following ten months, with AIM-7s being

fired alongside the AIM-9s. Additional trials utilizing the M61A1 rotary cannon also were conducted.

As placement of the M61A1 in the nose of the F/A-18 had been questioned by Navy analysts, its trials proved of more than passing interest. Accordingly, the 570 round magazine was expended during one test in one continuous firing run, and in others in six short bursts (which was more representative of the combat norm). As gun gas exhaust patterns were considered critical to both engine operation and pilot visibility, the tests were thoroughly analyzed, and found to prove the soundness of MCAIR's design decision.

In its ground support role, the F/A-18 was optimized to carry an extraordinary load. MCAIR proved the aircraft's ability to meet spec when a test mission, utilizing BuNo. 161248, was completed successfully with the aircraft having departed PAX River for the Pinecastle Range Complex in Florida with the following load: 4 x Mk.83 1,000 lb. bombs; 2 x AIM-9s, 3 x 315 gal. drop tanks; a Martin Marietta laser spot tracker/Perkin-Elmer strike camera (LST/SCAM) pod; a Ford FLIR pod; and a full complement of 570 20mm rounds for the M61A1 cannon. With this load, the aircraft had a gross takeoff weight of 48,253 lbs. The mission was completed in just over 3 hours and upon landing, sufficient fuel remained for nearly another hour of flight.

During mid- and late-1980, climatic testing of the F/A-18 was carried out under the auspices of the AF's 3246th Test Wing's McKinley Climatic Laboratory at Eglin AFB, Florida. Temperature extremes varying between – 65° F and + 125° F were generated, along with simulated high winds of up to 100 mph and rains of up to 20 in. per hour.

During the weapons and external stores trials, several changes were found to be necessary. Included were the moving of the external underwing stores racks forward some 5 in. in order to alleviate a rack flutter problem, and a redesign of the drop tanks. The latter came about as a result of fatigue anomalies that developed during carrier compatibility trials. The original elliptical cross section/spun-fiber-impregnated-with-aluminum tanks were replaced with conventional round aluminum ones, which consequently provided an additional 15 gals. of fuel.

Further F/A-18 test projects undertaken and completed by the FSD aircraft included maintenance engineering inspections, update and modification programs under the Hornet Hustle program, and a detailed review of the aircraft's radar cross section signature and electromagnetic

spectrum compatibility profile. Calculated carrier deck radiation levels pertaining to the latter were 200V/m (which compares to approximately 200V/m for the General Dynamics F-16, and 2V/m for earlier aircraft configurations). Some failings eventually were uncovered during these studies but improved electrical filters and increased component hardening eliminated them.

During November, ongoing spin and high angle-of-attack (AoA) profile flights (79° AoA w/13° yaw; 74° AoA w/25° yaw; and 65° AoA w/a sideslip angle of – 15° to + 30° were accomplished) were conducted utilizing BuNo. 160780. This aircraft, modified to incorporate an externally-mounted anti-spin chute on the empennage topside, was lost on November 14, when Lt. C. T. Brannon of VX-4 determined the aircraft to be out of control following spin initiation at 20,000 ft.

Brannon's accident served to alert MCAIR and Navy officials to a rare but deadly spin anomaly and no less than 110 spin test missions were conducted by MCAIR and Navy pilots in order to discover the cause and cure. though the cause of a spin could be any one or a number of control or flight envelope excesses, recovery could be accommodated by the judicious application of asymmetric thrust and associated control inputs. Proper and timely use of these two options effectively neutralized the out-of-control condition. Accordingly, computer software was developed to assist the pilot during the rare instances when more than 15° of yaw per second were encountered (i.e., as a spin was entered).

Full-scale development testing, by late 1980, now essentially was complete. The F/A-18 had cleared most of its performance hurdles without major trauma and MCAIR and Navy test pilots had thoroughly explored its flight envelope while developing confidence in it as a fully capable combat aircraft.

OPERATIONAL SERVICE:

The first Navy fleet readiness squadron to receive the F/A-18 was VFA-125 (Rough Raiders) at NAS Lemoore, California, which was commissioned as a mixed Navy/Marine Corps F/A-18 squadron on November 13, 1980. The first VFA-125 aircraft, out of the initial pilot production F/A-18 series, was delivered on February 19, 1981. The first full-up production F/A-18A arrived during September, and by the end of the year, a total of eight aircraft were on hand. This unit effectively was charged with training both ground and air crew for the Navy and Marine Corps' operational units and as such, was the

F/A-18A, BuNo. 162467, of VMFA-531, on October 11, 1986 during a transient stop at Carswell AFB. Squadron markings, including skull and tail code lettering, are in black. All other markings are standard for type.

F/A-18A of VMFA-531. Aircraft is equipped with three AIM-7E "Sparrows" and four AIM-9L "Sidewinders". Asymmetry of load apparently has little affect on "Hornet" performance. Tail markings are gray on gray, though "01" on nose is in black.

F/A-18A, BuNo. 163119 of VFA-15 at Cecil Field on April 22, 1987. Vertical fin stripe and miscellaneous identifiers are black. Tail code and lion caricature are in white. All other markings, including low-visibility insigne, are standard for type.

F/A-18A, BuNo. 163122 of VFA-15 transient at Luke AFB on May 9, 1987. White (with beige nose cone and black "Valions" lettering) centerline tank is noteworthy. All other markings are standard for a VFA-15 F/A-18A.

F/A-18A, BuNo. 161942 of VFA-25. Markings on tail are light gray on gray. "VFA-25" on fuselage side is in white. "403" on nose and "03" on vertical fin tip are in black. All other markings are standard for type.

F/A-18As of VFA-25. At the time, the aircraft were assigned to the USS "Constellation". Markings were the same as those found on BuNo. 161942. Slight flap displacement, which is standard in cruise configuration, is noteworthy.

F/A-18A, BuNo. 161944 of VFA-25 on September 22, 1984. Drooped flaperon position is typical of the aircraft when hydraulic power is off (i.e., whenever the engine is not running). Small VFA-25 logo on centerline tank nose is noteworthy.

F/A-18A, BuNo. 161957 of VFA-25. Aircraft is carrying a single Mk.82 500 lb. "slick" iron bomb on its starboard outer wing pylon. USS "Constellation" logo in white is barely visible near the tip of the vertical fin.

"HARM"-equipped F/A-18C, BuNo. 163453 of VFA-86 at Scott AFB on June 18, 1988. This aircraft represents both the newest "Hornet" configuration to enter the Navy inventory and one of the newest "Hornet" squadrons.

F/A-18A, BuNo. 163107 of VFA-87 during transient stopover at NAS Dallas on February 14, 1988. Markings such as code and indian head are in black. Aircraft mounts a single AGM-62 "Walleye" air-to-surface missile on its port outboard wing pylon.

pioneer squadron in terms of F/A-18 service. Because of this status, VFA-125's *Hornet* complement was larger than normal, at 60 aircraft, including a sizable percentage of F/A-18B trainers. Initially it was tasked with converting operational squadrons to the F/A-18 at a rate of some four per year.

Following Navy Bureau of Inspection and Survey trials during the early part of the year, VFA-125 served to finalize the F/A-18's training syllabus during a flight program lasting from September 27 to October 4, 1982, when it completed the first carrier qualification operations during a tour aboard the U.S.S. *Constellation*. Six VFA-125 pilots flying F/A-18Bs flew missions from the carrier, accumulating 57 day and 24 night traps as well as ten bolters.

The F/A-18 was declared operational on January 7, 1983, with Marine Fighter/Attack Squadron 314 at MCAS El Toro, California (by which time the *Hornet* fleet had logged more than 16,500 flight hours and 12,000 sorties in test and training flights), and later with VMFA-323 and VMFA-531. During April 1984, a second fleet readiness squadron, VFA-106 (*Gladiators*) was commissioned at NAS Cecil Field, Florida. A third (Marine), VMFAT-101 (*Sharpshooters*), followed at MCAS El Toro during 1988.

On February 1, 1985, the first Atlantic Fleet F/A-18 operational squadrons began forming at NAS Cecil Field, Florida, after undergoing conversion at NAS Lemoore, California. Also during that month, two west coast F/A-18 squadrons, VFA-113 *Stingers* and VFA-25 *Fist of the Fleet*, were deployed aboard the USS *Constellation* for the aircraft's first extended sea duty (to the Indian Ocean).

By June 1988, the following U.S. Marine Corps and U.S. Navy squadrons were F/A-18 equipped, or due for conversion into the type:

VMFAT-101 *Sharpshooters* (transitioned from F-4s during 1987/became operational at MCAS El Toro during 3/88); tail coded SH; initial assignment MAG-11, 3rd MAW as second West Coast F/A-18 Training Squadron.

VMFA-115 *Silver Eagles* (transitioned from F-4s at NAS Lemoore/became operational at MCAS Beaufort during 6/85); tail coded VE; initial assignment MAG-31, 2nd MAW.

VMFA-122 *Crusaders* (transitioned from F-4s and stood up at MCAS Beaufort during 4/86); tail coded DC; initial assignment MAG-31, 2nd MAW.

VMFA-212 *Lancers* (to be transitioned from F-4s at MCAS El Toro during 10/88; tail coded WD; will be assigned to MAG-24, 1st Brigade at MCAS Kaneohe.

VMFA-232 *Red Devils* (to be transitioned from F-4s at MCAS El Toro during 1989); tail coded WT; will be assigned to MAG-24, 1st Brigade at MCAS Kaneohe.

VMFA-235 *Death Angels* (to be transitioned from F-4s at MCAS El Toro during 1989); tail coded DB; will be assigned to MAG-24, 1st Brigade at MCAS Kaneohe.

VMFA-251 *Thunderbolts* (transitioned from F-4s at NAS Cecil Field and stood up at MCAS Beaufort during 6/86); tail coded DW; initial assignment MAG-31, 2nd MAW.

VMFA-312 *Checkerboards* (transitioned from F-4s at NAS Cecil Field during 87/became operational at MCAS Beaufort during 3/88); tail coded DR; initial assignment MAG-31, 2nd MAW.

VMFA-314 *Black Knights* (transitioned from F-4s at NAS Lemoore during 1982/became operational at MCAS El Toro during 1/83); tail coded VW; initial assignment MAG-11, 3rd MAW.

VMFA-323 *Death Rattlers* (transitioned from F-4s at NAS Lemoore during 1983/became operational at El Toro during 3/83); tail coded WS; initial assignment MAG-11, 3rd MAW.

VMFA-333 *Shamrocks* (transitioned from F-4s at NAS Cecil Field during 1987/became operational at MCAS Beaufort during 8/87); tail coded DN; initial assignment MAG-31, 2nd MAW.

VMFA-451 *Warlords* (transitioned from F-4s at NAS Cecil Field during 1987/became operational at MCAS Beaufort during 6/87); tail coded VM; initial assignment MAG-31, 2nd MAW.

VMFA-531 *Grey Ghosts* (transitioned from F-4s at NAS Lemoore during 1983/became operational at MCAS El Toro during 6/83); tail coded EC; initial assignment MAG-11, 3rd MAW.

VFA-15 *Valions* (transitioned from A-7s at NAS Cecil Field during 1986/became operational there during 4/87); tail coded AC; initial carrier assignment CVW-3 aboard CVN-71.

VFA-25 *Fist of the Fleet* (transitioned from A-7s at NAS Lemoore during 1983/became operational there during 12/83); tail coded NK; initial carrier assignment CVW-14 aboard CV-64.

VFA-81 *Sunliners* (transitioned from A-7s at NAS Cecil Field during 1987/became operational there during 5/88); tail coded AA; initial carrier assignment CVW-17 aboard CV-60.

VFA-82 *Marauders* (transitioned from A-7s at NAS Cecil Field during 1987/became operational there during 12/87; tail coded AJ; initial carrier assignment CVW-8 aboard CV-66.

VFA-83 *Rampagers* (transitioned from A-7s at NAS Cecil Field during 1988/became operational there during 6/88); tail coded AA; initial carrier assignment CVW-17 aboard CV-60.

VFA-86 *Sidewinders* (transitioned from A-7s at NAS Cecil Field during 1987/became operational there during 3/88); tail coded AJ; initial carrier assignment CVW-8 aboard CV-66.

VFA-87 *Golden Warriors* (transitioned from A-7s at NAS Cecil Field during 1986/became operational there during 12/86); tail coded AC; initial carrier assignment CVW-3 aboard CVN-71.

VFA-106 *Gladiators* (commissioned at NAS Cecil Field on 4/27/84/became operational there during 4/85); tail coded AD; East Coast Training Squadron.

VFA-113 *Stingers* (transitioned from A-7s at NAS Lemoore during 1983/became operational there during 9/83); tail coded NK; initial carrier assignment CVW-14 aboard CV-64.

VFA-125 *Rough Riders* (commissioned at NAS Lemoore on 11/13/80/became operational there during 2/81); tail coded NJ; West Coast Training Squadron.

VFA-131 *Wildcats* (commissioned at NAS Lemoore on 10/3/83/became operational there during 7/84, then transferred to NAS Cecil Field); tail coded AK; initial carrier assignment CVW-13 aboard CV-43.

VFA-132 *Privateers* (commissioned at NAS Lemoore on 1/3/84/became operational there during 10/84, then transferred to NAS Cecil Field); tail coded AK; initial carrier assignment CVW-13 aboard CV-43.

VFA-136 *Knighthawks* (commissioned at NAS Lemoore on 7/1/85/became operational there during 4/86 and then transferred to NAS Cecil Field); tail coded AK; initial carrier assignment CVW-13 aboard CV-43.

VFA-137 *Kestrels* (commissioned on 7/1/85 at NAS Cecil Field/became operational there during 4/86); tail coded AK; initial carrier assignment CVW-13 aboard CV-43.

VFA-146 *Blue Diamonds* (scheduled to transition from A-7s at NAS Lemoore during 1989).

VFA-147 *Argonauts* (scheduled to transition from A-7s at NAS Lemoore during 1989).

VFA-151 *Vigilantes* (transitioned from F-4s at NAS Lemoore during 1986/became operational there during 10/86); tail coded NF; initial carrier assignment CVW-5 aboard CV-41 (home base NAF Atsugi).

VFA-161 *Chargers* (transitioned from F-4s at NAS Lemoore during 1986/became operational at NAS Lemoore during 10/86; decommissioned 3/88); tail coded NM; initial carrier assignment CVW-10 but decommissioned before being sent to sea (originally scheduled for *Independence*).

VFA-192 *Golden Dragons* (transitioned from A-7s at NAS Lemoore during 1986/became operational there during 7/86); tail coded NF; initial carrier assignment CVW-5 aboard CV-41 (home base NAF Atsugi).

VFA-195 *Dambusters* (transitioned from A-7s at NAS Lemoore during 1985/became operational there during 1/86); tail coded NF; initial carrier assignment CVW-5 aboard CV-41 (home base NAF Atsugi).

VFA-203 *Blue Dolphins* (scheduled to complete transition from A-7s at NAS Cecil Field during 3/91); tail coded AF; initial assignment CVW-20; third reserve unit.

VFA-303 *Golden Hawks* (transitioned from A-7s at NAS Lemoore during 1984/became operational there during 9/85); tail coded ND; initial assignment CVW-30; first reserve unit.

VFA-305 *Lobos* (transitioned from A-7s at NAS Lemoore/became operational at Pt. Mugu during 1/88); tail coded ND; initial assignment CVW-30; second reserve unit.

VX-4 *Evaluators* (test work begun at NAS Point Mugu during 1981); tail coded XV.

VX-5 *Vampires* (test work begun at NWC China Lake during 1981); tail coded XE.

Strike Warfare Center (received two F/A-18As during 1985); no tail code (lightning bolt).

Strike Aircraft Test Directorate (received F/A-18 complement at NATC Patuxent River during 1979); no tail code.

Naval Weapons Center (received three test F/A-18s, BuNos. 161366, 161713, 161720 during 1979); NWC eagle on tail.

Pacific Missile Test Center (received two test F/A-18s for AMRAAM testing during 1985); PMTC insignia on tail.

NASA (Ames/Dryden Flight Research Center received the first of nine FSD F/A-18s, BuNos. 160775 [non-flying/for parts], 160777, 160778, 160780, 160781, 160782, 160785, 161216, 161250, and 161251 during 1985).

Navy Test Pilots School (received F/A-18B BuNo. 161356 and F/A-18B BuNo. 161249; NTPS on tail.

The following squadrons presently are planned to become F/A-18 squadrons as of the year indicated in parenthesis: VMFA-112 (1994; Dallas/Reserves); VMFA-134 (1989; El Toro/Reserves); VMFA-142 (1993; Cecil/Reserves); VMFA-321 (1992; Washington/Reserves); VMFA-322 (1995; Dallas/Reserves); VFA-22 (1992; Lemoore); VFA-27 (1991; Lemoore); VFA-37 (1991; Cecil); VFA-47 (1991; Cecil); VFA-72 (1991; Cecil); VFA-94 (1992; Lemoore); VFA-97 (1991; Lemoore); VFA-105 (1991; Cecil); and VFA-204 (1991; New Orleans/Reserves).

It should be noted that a total of thirteen active Marine Corps fighter/attack squadrons presently are planned, based on the 316 Marine F/A-18s on order. Marine Corps tail codes, as a point of interest, do not follow the standard pattern of Navy Atlantic and Pacific Carrier Air Wings, though VMFA-314's and VMFA-323's VS and WS codes were replaced by the AK code of the Navy's CVW-13 (U.S.S. *Coral Sea*) during the course of their deployment.

It also should be noted that the following aircraft carriers are presently the only ones currently considered "F/A-18 capable": CV-41 *Midway*, CV-43 *Coral Sea*, CV-60 *Saratoga*, CV-62 *Independence*, CV-64 *Constellation*, CV-66 *America*, and CVN-71 *Theodore Roosevelt*. The following aircraft carriers are scheduled to become F/A-18 capable during the coming decade: CV-59 *Forrestal*, CV-61 *Ranger*, CV-63 *Kitty Hawk*, CVN-65 *Enterprise*, CV-67 *John F. Kennedy*, CVN-68 *Nimitz*, CVN-69 *Eisenhower*, CVN-70 *Carl Vinson*, CVN-72 *Abraham Lincoln*, and CVN-73 *George Washington*.

The only other U.S. Navy unit currently operating the F/A-18 is the service's prestigious *Blue Angels* precision aerobatic team. During February 1986, the F/A-18 formally was selected to replace the team's aging Douglas A-4F *Skyhawks* and accordingly, eight F/A-18As (BuNos. 161520, 161521, 161523, 161524, 161525, 161526, 161527, and 161588) and a single F/A-18B (BuNo. 161355), representing early production aircraft no longer considered suitable for shipborne operation, were modified to incorporate smoke-generating systems, a revised and aerobatic-routine-optimized flight control system, special seat harnesses, a VOR navigation unit, and a civilian ILS in place of the military ACLS. Additionally, the nose-mounted M61 gun was removed (all other extant weapons system related equipment, including the Hughes manufactured AN/APG-65 radar was left in place). The aircraft were painted in standard *Blue Angels* blue, yellow, and gold markings.

It should be noted that *Blue Angels'* tail numbers (i.e., 1 through 7) often are switched between aircraft depending on their respective maintenance statuses. Out the eight single-seat aircraft currently available, two usually are kept on "spares" status.

As of early 1988, the status of the Navy/Marine Corps F/A-18 program included contracts for 1,007 F/A-18As and approximately 150 F/A-18Bs. Included in the 1,157 total aircraft were the 11 FSD airframes. The 500th *Hornet*, an F/A-18A, BuNo. 163133, was delivered to VMFA-451 at MCAS Beaufort on April 8, 1987. Production rates (in parenthesis) on a yearly basis have been as follows: 1979 (9); 1980 (25); 1981 (60); 1982 (63); 1983 (84); 1984 (84); 1985 (84); 1986 (84); and 1988 (84). Production is scheduled to be reduced to 72 per year from 1989 onwards.

By early 1987, Navy and Marine Corps F/A-18A/Bs had logged no less than 264,086 flight hours. Additionally, the type's first incursion into actual combat occurred during April 1986 when VFA-131, VFA-132, VMFA-314, and VMFA-323 aircraft participated in retaliatory missions against Libyan targets from the U.S.S. *Coral Sea*. Both prior to and during the attacks, F/A-18s were utilized to suppress SAM sites with HARM anti-radiation missiles. Other missions included airborne CAPS. The F/A-18s generated approximately 80% of the deck launched intercepts undertaken during the course of this short-term conflict.

FOREIGN SERVICE:

Canada

During March 1977, the Canadian government authorized its Department of National Defense to solicit proposals from industry for a New Fighter Aircraft (NFA) to replace its rapidly aging fleet of Lockheed CF-104 *Starfighters*, McDonnell CF-101 *Voodoos*, and Northrop CF-5 *Freedom Fighters*. Some 130 to 150 aircraft were needed and it was recommended the aircraft be (1) "off-the-shelf" with a minimum of Canadian-unique features; (2) purchased within a fixed-price budget of $2.34 billion 1977 Canadian dollars; and (3) capable of providing con-

F/A-18A, BuNo. 163106 of VFA-87 transient at Andrews AFB on September 4, 1987. Markings are conventional for type, with the exception of the squadron identifier on the centerline drop tank. Placement of tail code is of interest.

F/A-18A, BuNo. 161969 of VFA-106. Aircraft mounts a triple-ejector-rack on its inboard wing pylon, but no weapons. Markings are standard gray on gray with the exception of "321" in black on the nose and drooped flaperon.

F/A-18B, BuNo. 163110 of VFA-106 at Cecil Field on April 22, 1987. Chevron and "AD" code on vertical tail surfaces, "300" on nose, flaperon, and vertical fin tip, and "VFA-106" on fuselage side are in black. All other markings are gray on gray.

F/A-18A, BuNo. 161710 of VFA-113 at NAS Lemoore on October 20, 1983. Markings show the early light gray on gray camouflage pattern with dark gray mascot and miscellaneous identifiers. "302" on nose is very dark gray.

siderable offset benefits to Canadian industry.

During September 1977, an RFP was issued to six contractors with a total of seven aircraft types being examined. A detailed evaluation of these was conducted by the NFA Project Office over the following year and during November 1978, the contenders were narrowed to the General Dynamics F-16 and the McDonnell Douglas F/A-18. A lengthy analysis of the two finalists now followed, with the F/A-18 being chosen the winner on April 10, 1980.

The initial contract called for 113 CF-18As and 24 CF-18Bs. Following the first flight of the first production example on July 29, 1982, deliveries utilzing CF-18Bs 188901 and 188902 were initiated on October 25. Though initially fraught with difficulties resulting from unpredicted premature vertical surface and fuselage section structural assembly fatigue and a temporary shortage of spare parts, delivery rates eventually reached predicted values of approximately two aircraft per month following a year of intermittent activity. The final delivery now is scheduled to take place during the early fall of 1988. The aircraft are assigned to North American Air Defense (NORAD) and North Atlantic Treaty Organization (NATO) defense duties.

Changes incorporated into the Canadian Air Force CF-18s differentiating them from their Navy/Marine Corps counterparts include a different instrument landing system, the addition of a spotlight on the left side of the fuselage for use during night intercepts of unidentified aircraft, and provision for carrying LAU-5003 rocket pods.

The following is a complete listing of all presently extant RCAF CF-18 squadrons:

Sqdn.	Base	# A/C	Role	Activated	Name
410 OTS	Cold Lake	23	training	6/82	Cougar
425 TFS	Bagotville	12	air def/ grnd attk	4/85	Alouette
409 TFS	Baden-Soellingen	16	grnd attk/ air def	6/85	Nighthawk
439 TFS	Baden-Soellingen	16	grnd attk/ air def	12/85	Tiger
421 TFS	Baden-Soellingen	16	grnd attk/ air def	6/86	Red Indians
441 TFS	Cold Lake	12	air def/ grnd attk	7/87	Silver Fox
433 TFS	Bagotville	12	grnd attk/ air def	12/87	Porcupine
416 TFS	Cold Lake	12	grnd attk/ air def	12/88	Lynx

The CAF operational vision eventually foresees the CF-18 at five forward locations in the Canadian far north. Included are Inuvik, Yellowknife, Iqaluite, and Rankin Inlet in the Northwest Territories, as well as Kuujjuaq, Quebec. Additionally, three squadrons, 410, 416, and 433 serve as NATO reinforcement squadrons under the auspices of the 1 Canadian Air Group (soon to be upgraded to a Canadian air division) in West Germany.

CAF CF-18 serial numbers are as follows: CF-18A, 188701 through 188813, inclusive; CF-18B, 188901 through 188924 inclusive.

Australia

The Australian government, via the Royal Australian Air Force, on October 20, 1981, made formal their announced plans to procure the F/A-18A/B for RAAF service. A total of 75 aircraft (57 F/A-18As and 18 F/A-18Bs) were to be bought, with the first two to be manufactured outright by McDonnell Douglas, and the remainder to be assembled initially from acquired MCAIR parts and later, from indigenously manufactured parts by Aerospace Technologies of Australia at their Avalon facility.

The two MCAIR-assembled aircraft, both F/A-18Bs, were delivered from St. Louis to NAS Lemoore and from there, by air, to RAAF Williamtown, near Sydney on May 17, 1985. The first Australian-assembled aircraft made its first flight on February 26, 1985 and was turned over to the RAAF's No. 2 OCU on May 4. On June 3, 1985, the first all-Australian F/A-18 (F/A-18B, A21-104) made its first flight.

Two RAAF squadrons, in addition to No. 2 OCU, now are operational with the F/A-18. These include No. 3 Squadron at Williamtown (which received its first F/A-18As, A21-8 and A21-9, during the autumn of 1986), and No. 77 Squadron (receiving its first F/A-18As during July 1987) also at Williamtown. No. 75 Squadron, the third, continues to transition at Williamtown as of this writing and will be stationed at RAAF Tindal NWT.

The RAAF's F/A-18s are tasked with air defense, anti-shipping/strike, and battlefield support. These missions are being taken-over from Dassault Mirage IIIOs. At present, the last RAAF F/A-18 is scheduled for delivery during May, 1990. RAAF F/A-18 serial numbers are A21-101 to A21-175, inclusive.

Spain

During May 1983, following a competition involving several other advanced state-of-the-art fighters, the Spanish government, via the Spanish Air Force, announced its plans to order 72 EF-18s (60 single seat and 12 two-seat; the original order was to have been for 144 aircraft, but budgetary constraints led to a halving of the original quantity) with an option to acquire 12 more as needed. The first C.15 (as the single-seat configuration is designated in Spain; the C stands for Caza—"fighter") was rolled out on November 22, 1985, and by October of the following year, nine aircraft had been delivered.

The initial C.15s and CE.15s (as the two-seat configuration is designated in Spain; CE stands for Caza de Entrenamiento—"fighter trainer") were used to train the first ten Spanish Air Force pilots in the U.S. They then

Two F/A-18As, BuNos. 161936 and 161926 bearing the standard gray on gray markings of VFA-113 depart Offutt AFB following a transient stopover on February 3, 1984. Gear retraction sequence of the aft aircraft is noteworthy, as is flap and flaperon position of both aircraft during takeoff.

F/A-18A, BuNo. 161353 in standard markings of VFA-125. Drooped flaperon is standard for the aircraft without powered hydraulic system. Aircraft is clean and seen without external stores of any kind. Up to 17,000 lbs. can be carried externally.

F/A-18B, BuNo. 161354 of VFA-125 during a transient stopover at Offutt AFB on July 11, 1982. High contrast gray on gray markings have since been subdued in subsequent schemes. Playboy bunny on nose of centerline tank is noteworthy.

F/A-18A, BuNo. 161972 of VFA-131 during a transient stopover at Buckley ANGB on December 22, 1984. This aircraft was the 190th ''Hornet'' manufactured by MCAIR. USS ''Coral Sea'' is discernible near the top of the vertical fin.

F/A-18A, BuNo. 161970 of VFA-131 transient at Offutt AFB. ''Wildcats'' artwork on vertical fin is in dark gray, as is USS ''Coral Sea'' near fin tip. Centerline drop tank is typically painted off-white to gray with a beige nose cap.

were delivered by their own crews to Spanish soil. An Operational Conversion Unit (15th Groupo) was created during 1986 following their arrival, and this served to accommodate conversion needs in-country.

Two wings and four squadrons are equipped with the C.15/CE.15. The wings are 15 Wing at Zaragoza and 12 Wing at Torrejon. Their respective squadrons are Escuadron 151 and Escuadron 152, and Escuadron 121 and Escuadron 122.

Spanish Air Force F/A-18 serial numbers are as follows: C.15-13 through C.15-72, inclusive; CE.15-1 through CE.15-12, inclusive.

Kuwait

Following a competition pitting the F/A-18 against the General Dynamics F-16 and the Dassault *Mirage 2000*, Kuwait became the first Arab nation to opt for the *Hornet* when, during early May 1988, it was announced that the U.S. Department of Defense would be presented a formal request for 40 aircraft. Congressional approval of the sale is expected, though some concern remains over technology transfer. Delivery dates and schedules have not yet been firmed up, according to MCAIR.

Miscellaneous

As with any major U.S. fighter program, foreign interest in the F/A-18 has remained strong, though relatively noncommital, since the advent of the aircraft. Serious inquiries have been made by Switzerland, South Korea (which continues to entertain the possibility of an F/A-18 acquisition and which has sent a team to the U.S. on several occasions for flight evaluation), and the United Arab Emirates. Other countries that have expressed lessserious interest in the F/A-18 include Israel, Japan, New Zealand, and Singapore.

Most recently, the French have begun negotiations with the Navy for the possible use of two F/A-18s in order to explore carrier suitability aboard the French carrier *Foch* (and forthcoming *Charles De Gaulle*). France is considering an interim buy of F/A-18s to replace its rapidly aging fleet of Vought F-8Ns until the introduction into service of its advanced *Rafale* fighter.

F/A-18 VERSIONS:

F/A-18A—The standard, initial production single-seat configuration. Optimized to serve with the Navy and Marine Corps as a replacement for the McDonnell F-4 and Vought A-7. Capable of carrying AIM-7 and AIM-9 air-to-air missiles, as well as miscellaneous air-to-surface weapons such as HARM and *Maverick*. Can be equipped with FLIR and laser tracker.

F/A-18B—The standard, initial production tandem two-seat configuration. Equipped with a full control complement in the back seat, including all necessary instrumentation. Optimized for training, but retains full combat capability. Addition of back seat forced an approximate 6% reduction in fuel capacity.

F/A-18C—An improved and updated single-seat configuration purchased with FY 86 funds and now the current production standard single-seat aircraft. Similar externally to its predecessors with the exception of miscellaneous antenna and antenna fairings and other details. Capability improvements include: the ability to carry up to ten AIM-120 advanced medium range air-to-air missiles (AMRAAM) and two fuselage stations and four wing pylons (two missiles on each pylon, max.; additionally, tests have been conducted with a single AMRAAM mounted on a special fuselage centerline pylon, giving the aircraft a total AMRAAM capability of 11; AMRAAM capability exists through a modified stick

grip, a new stores management processor, and hardware changes in the launchers and AN/APG-65 radar); the ability to carry up to four imaging infrared *Maverick* air-to-surface missiles on four wing pylons; provision for the AN/ALQ-165 airborne self-protection jamming (ASPJ) system (which permits interchangeability with the AN/ALQ-126B countermeasures system) and 11 new or updated antennas, and consequently, an improved environmental control system to accommodate the additional cooling requirement (ECP-35); common nose provisions for reconnaissance equipment that include embedded hard points, electrical wiring, ECS provision, and EMI shielding; an upgraded stores management set with 128K memory; an Intel 8086 processor; an upgraded armament multiplex bus to meet MIL-STD 1553B and MIL-STD 1760 weapons interface capability; a flight incident recorder and monitoring set (FIRAMS) which incorporates an integrated fuel/engine indicator (ECP-178); a data storage set for recording maintenance and flight incidents data; a signal data processor interfacing with the fuel system to provide overall system control; an independent l/r motive flow fuel system (ECP-162); enhanced built-in-test capability and automatic adjustment of aircraft c.g. as fuel is depleted; a maintenance status panel enabling avionics faults to be isolated at circuit card level; a new XN-6 mission computer with higher processing speeds and double the memory of its XN-5 predecessor; and an advanced Martin-Baker NACES (Navy Aircrew Common Ejection Seat) ejection seat.

The first F/A-18C, piloted by MCAIR company test pilot Glen Larson, made its first flight at MCAIR's St. Louis facility on September 3, 1986 and was delivered to NATC Patuxent River on September 21. The first F/A-18C to be formally delivered to the Navy (thought to be production verification ship #247) arrived at the Naval Weapons

F/A-18A of VFA-132 during a 1985 airshow at NAS Dallas. Tail code on the inside of the vertical fin is medium to dark gray on gray. Only distinctive identifier is the ''210'' seen on the aircraft's nose.

F/A-18A, BuNo. 162437 of VFA-132 transient at Carswell AFB on October 30, 1986. Mottled appearance of camouflage is due to weathering. ''205s'' on nose and vertical fin tip are in black. Nose cap of centerline tank is graphite gray.

F/A-18A, BuNo. 162845 of VFA-136 on August 8, 1986. Tail markings, including eagle, USS "Coral Sea", and miscellaneous identifiers all are in black. All other markings are standard for type.

F/A-18A, BuNo. 162854 of VFA-137 at Cecil Field on April 22, 1987. Aircraft is carrying a 1,000 lb. Mk.83 "slick" iron bomb on a horizontal ejector rack attached to its outboard wing pylon and a 330 gal. external wing tank on its inboard.

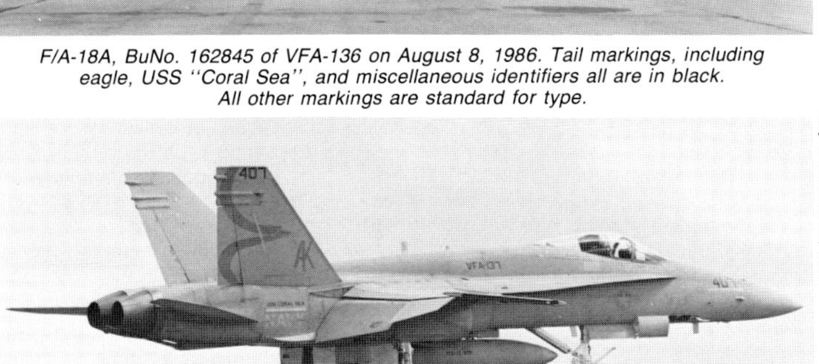

F/A-18A, BuNo. 162861 of VFA-137 on final approach to NAS Dallas during 1987. Flap and flaperon extension angle is noteworthy, as are extended struts of main landing gear. Notation in black on centerline tank is VFA-137 NFWS.

F/A-18A, BuNo. 162888 of VFA-151 transient at NAF Andrews on September 6, 1986. Markings are typical gray on gray, with "201" identifiers in black. Noteworthy is squadron identifier and badge in black on nose of centerline fuel tank.

Center, China Lake on September 23 piloted by LCDR John Bell.

Flight testing of these and several other F/A-18Cs continues as of this writing, with several aircraft configured to undertake night attack capability trials. In this configuraton the pilot will be equipped with night vision goggles and the aircraft will have dedicated cockpit lighting and an externally-mounted AN/AAR-50 TINS (Thermal Imaging Navigation Set—due for delivery durig early fall 1989) pod, a raster-type head-up display, multi-function color displays, and a digital mapping system. As a point of interest, all F/A-18C/Ds delivered from October 1989-on will be night attack capability equipped.

F/A-18D—An improved and updated tandem two-seat configuration purchased with FY 86 funds and now the current production standard two-seat aircraft. The aft cockpit remains fully mission capable but with independent display selection. An advanced F/A-18D configuration, known as the F/A-18D+, and featuring improved and multi-disciplined "uncoupled cockpits" presently is being considered for the Marine Corps.

F/A-18(R)—Evaluation of design proposals from MCAIR calling for a simple reconnaissance conversion to the stock F/A-18A (and later, F/A-18C) was undertaken by the Navy during the Autumn of 1982. After concluding that the recce modification was feasible and acceptable to Navy requirements, a single aircraft, F/A-18A BuNo. 161214, was set aside for modification and flight test. This aircraft successfully undertook the first F/A-18(R) flight on August 15, 1984.

The reconnaissance system package is designed to fit in the space normally occupied by the aircraft's M61 gun (see above under F/A-18C). Lower nose fairings are removed and replaced with a hinged and slightly bulged hatch containing optical transparencies for cameras. The latter consist of a Fairchild-Weston KA-99 low/medium-altitude panoramic unit and/or a Honeywell AN/AAD-5 linescan unit. Additional sensors, including a low altitude, high-speed, high-resolution camera have been studied.

The entire reconnaissance package is designed to be removable and easily replaced by the conventional gun system, thus giving the aircraft a dual mission capability on short notice. Emphasis now is being placed on digitized, real-time transmittable data gathering sensor systems.

RF-18D—This configuration entails internal systems modifications only (see above under F/A-18C), as the actual sensor gear is carried in an all-weather reconnaissance pod that will be mounted on the centerline stores station beneath the aircraft fuselage. The actual pod, weighing approximately 1,500 lbs. and under development by the Loral Corporation's Defense Systems Division, contains an improved version of the AN/UPD-4 high-resolution synthetic aperture side-looking radar called the AN/UPD-9. This unit, with a range of approximately 50 miles and developed specifically in response to a Marine Corps request, when coupled with the aforementioned optical and IR linescan sensors that can be mounted in the aircraft's nose, will provide a versatile and advanced reconnaissance package to replace the systems currently in use. Imagery from the SLAR will be capable of being transmitted in real-time over a range of just over 200 miles via a datalink, but can be displayed in the RF-18D's rear cockpit, as well. Service entry of the RF-18D is expected to occur sometime during 1990.

Advanced pod configurations, containing electro-optical sensors with improved real-time datalink capability already are under development, and flight testing of these units is ongoing under the auspices of Loral and participating agencies.

CF-18A—F/A-18A optimized for use by the CAF. Information pertaining to select differences can be found in the Canadian section.

CF-18B—F/-18B optimized for use by the CAF. Information pertaining to select differences can be found in the Canadian section.

C.15—F/A-18A optimized for use by the Spanish Air Force. Information pertaining to select differences can be found in the Spanish section.

CE.15—F/A-18B optimized for use by the Spanish Air Force. Information pertaining to select differences can be found in the Spanish section.

Super Hornet—This is a generic label describing a series of advanced F/A-18 configuration studies optimized for service well into the 21st Century. This title also was used to refer to an upgraded F/A-18 development submitted to the Japan Air Self-Defence Force for their FSX requirement, and similarly to a configuration submitted to the Royal Air Force and West German Air Force as an alternative to the European Fighter Aircraft (EFA).

MCAIR has been decidedly reluctant to release details concerning the Super Hornet, but purportedly accurate configurations include designs with "cranked arrow" wings, canards, new vertical and horizontal tail surfaces, CCV technology, increased use of lightweight composite materials, and uprated engines. Subsequent to the major redesign profferings, a less radically modified F/A-18 configuration, offering improved engines, reconfigured and enlarged intakes, and a general aerodynamic clean-up also was under consideration by MCAIR and its major customers.

The Navy, in projecting its needs into the 21st Century, has examined closely four basic Hornet 2000 design options and apparently has settled on a compromise configuration that has a slightly larger wing, improved avionics and radar, increased survivability, electronic warfare suite improvements, an advanced cockpit with multi-sensor integration and helmet mounted sight/displays, airframe upgrades, significantly upgraded F404 engines, and a fuselage plug inserted behind the cockpit to provide additional space for fuel and avionics. Called the "3A", it has revised vertical stabilizers and other minor external modifications, but otherwise is similar over-all to the extent aircraft.

Near-term, future systems tentatively scheduled for integration into the F/A-18 weapons systems include an improved radar with new processors based on the APG-71, F-14D fire control system that will allow greater flexibility and speed plus new modes to include, interleaved air-to-air and air-to-surface modes; a radar beacon mode for offset bombing, coupled terrain following; high-resolution synthetic aperture maps; upgraded engines; more internal fuel; and various survivability enhancements (EW).

U.S. Navy and Marine Corps YF-17 and F/A-18 BuNos. (as of 6/88):

YF-17—72-1569/72-1570

F/A-18A—160775/160780, 160782/160783, 160785, 161213/161216, 161248, 161250/161251, 161353, 161358/161359, 161361/161367, 161519/161528, 161702/161710, 161712/161713, 161715/161718, 161720/161722, 161724/161726, 161728/161732, 161734/161739, 161741/161745, 161747/161761, 161925/161931, 161933/161937, 161939/161942, 161944/161946, 161948/161987, 161294/162406, 162423/162426, 162428/162477, 162826/162835, 162837/162841, 162843/162849, 162851/162856, 162858/162863, 162865/162869, 162871/162875, 162877/162884, 162886/162909, 163092/163103, 163105/163109, 163111/163114, 163116/163122, 163124/163175

F/A-18B—160781, 160784, 161217, 161249, 161354/161357, 161360, 161704, 161707, 161711, 161714, 161723, 161727, 161733, 161740, 161746, 161924, 161932, 161938, 161943, 161947, 162402, 162427, 162836, 162842, 162850, 162857, 162864, 162870, 162876, 162885, 163104, 163110, 163123

RF/A-18A—161214

F/A-18C—163427/163433, 163435, 163437/163440, 163442/163444, 163446, 163448/163451, 163453, 163455/163456, 163461/163463, 163465/163467, 163469/163741, 163473, 163475/163478, 163480/163481, 163483/163484, 163485, 163487, 163489/163491, 163493/163496, 163498/163499, 163502/163506, 163508/163509, 163699, 163701/163706, 163708/163719, 163721/163733, 163735/163748, 163750/163762, 163764/163770, 163772/163777, 163779/163782, 163985, 163987/163988, 163990, 163992/163993, 163995/163996, 163998/164000, 164002/164004, 164006/164008, 164010, 164012/164013, 164015/164016, 164018, 164020/164021, 164023, 164025, 164027, 164029/164031, 164033/

164034, 164036/164037, 164039, 164041/164042, 164044/164045, 164047/164048, 164050, 164052, 164054/164055, 164057, 164059/164060, 164062/164063, 164065/164067

F/A-18D—163434, 163436, 163441, 163445, 163447, 163452, 163454, 163457, 163460, 163464, 163468, 163472, 163474, 163479, 163482, 163486, 163488, 163492, 163497, 163500/163501, 163507, 163510, 163700, 163700, 163707, 163720, 163734, 163749, 163763, 163771, 163778, 163086, 163909, 163991, 163994, 163997, 164001, 164005, 164009, 164011, 164014, 164017, 164019, 164022, 164024, 164026, 164028, 164032, 164035, 164038, 164040, 164043, 164046, 164049, 164051, 164053, 164056, 164058, 164061, 164064, 164068

CONSTRUCTION AND SYSTEMS:

The F/A-18 is a single- or two-seat (tandem) multi-mission fighter capable of delivering both air-to-air and air-to-surface weapons. Construction materials are state-of-the-art with 40% of the aircraft's surface area made up of lightweight, corrosion- and fatigue-resistant carbon fiber/epoxy materials. Airframe materials in general consist of 50.2% aluminum alloy, 15.5% steel alloy, 13.2% titanium alloy, 10.1% carbon/epoxy composites, and 11% miscellaneous.

The following description is applicable primarily to the F/A-18C (and where appropriate, the F/A-18D), but also generally applicable to the F/A-18A (and where appropriate, the F/A-18B):

Cockpit: The pressurized cockpit (cockpit altitude remains constant at approximately 8,000 ft. to an altitude of 23,000 ft.; above 23,000 ft., cockpit altitude increases slowly to approximately 14,500 ft. at 35,000 ft., and 20,000 ft. at 50,000 ft.) accommodates one or two crew members in tandem on Martin-Baker SJU-6/A (front) or SJU-5/A (rear) NACES ejection seats (CF-18A/B seats are referred to as SJU-6A [modified] and SJU-5A [modified], respectively). The ejection seats are of the ballistic catapult/rocket type. The cockpit is accessed via a boarding ladder stowed under and integral to the left-side LEX. Ladder extension and retraction can only be accomplished from outside the aircraft. The cockpit is fully air-conditioned and heated (via engine compressor section bleed air), and equipped with a hinged, upward-opening single-piece canopy (both single- and two-seat aircraft) that when closed permits unobstructed 360° visibility. Canopy actuation is electro-mechanical.

The display and control technology found in the advanced cockpit is simplified so that a single crew member can easily handle the entire mission. The HUD and critical controls, located on the Hands-On-Throttle-And-Stick (HOTAS) controls, allow the pilot to fly the aircraft, find targets, and deliver weapons without ever having to take his eyes off the action or his hands off the flight controls. A windshield anti-ice and rain removal system is provided.

Fuselage: The fuselage is of semi-monocoque construction utilizing aluminum, steel, and composites in appropriate locations. A titanium firewall is mounted between the engine bays. The empennage supports the twin vertical tail surfaces and rudders, and a hydraulically-actuated composite construction hinged, upward-opening airbrake.

Wings: The wing is a conventional cantilever design placed approximately midway on the fuselage. Structural construction materials consist of aluminum alloys with graphite/epoxy inter-spar skin panels and trailing edge flaps. The multi-spar structure supports full-span leading edge maneuvering flaps with a maximum extension angle of 30°. Boundary layer control is achieved via slots in the wing root sections. Bertea hydraulic actuators move the single-slotted trailing edge flaps, giving them a maximum droop angle of 45°. Additionally, the outboard-mounted ailerons, actuated by Hydraulic Research rams, can be drooped to 45° in concert with the flaps. Both the leading and trailing edge flaps are computer programmed to droop for optimum lift and drag in both maneuvering and cruise conditions, and both the trailing edge flaps and ailerons deflect differentially for roll control. The LEX surfaces provide additional lifting surface area and generate beneficial vortices affecting directional stability and lift at angles of attack approaching 60°. The outer wing panels and ailerons, powered by Garrett AiResearch mechanical drives, fold upward at an angle of 100°.

Tail Surfaces: The tail surfaces consist of a pair of conventional single-piece slab stabilators and two verticals. All surfaces are cantilever and are constructed primarily of graphite/epoxy composites over a light alloy honeycomb core. The verticals cant outboard 20° from the vertical. The horizontal stabilators are of the all-moving variety with 2° anhedral built-in. They are actuated collectively and differentially by National Water Lift hydraulic rams.

Landing Gear: The tricycle landing gear is fully retractable and consists of a twin-wheel nose unit and two single-wheel main units. The landing gear is electrically controlled and hydraulically operated. Before the gear can be raised normally, the weight must be off the wheels and the launch bar must be retracted. When the gear is extended, all gear doors remain open. The nose unit retracts forward and the mains, aft (turning 90° to permit horizontal stowage inside the lower section of each intake tunnel assembly). Each main gear is equipped with Bendix wheels and disc brakes. The latter are equipped with an anti-skid system which, when activated, prevents brake application until 5 secs. after touchdown. The nosewheel tire size is 22 x 6.6-10, 20 ply, and each is pressurized to 350 psi for carrier ops and 150 psi for land ops. Mainwheel tire size is 30 x 11.5-14.5, 24 ply, and each is presurized to 350 psi for carrier ops and 200 psi for land ops. The Ozone nose wheel unit is equipped with a combination shimmy damper and dual mode steering system. It is electrically controlled by two switches on the stick grip in the cockpit, and hydromechanically operated through inputs from the rudder pedals and flight control computers. Steering is provided in a low mode (+ or − 16°) or a high mode (+ or − 75°). On the ground, nose wheel steering is disengaged when power is removed from the aircraft. The nose gear strut also is equipped with a conventional, retractable towbar to accommodate catapult launch requirements. It is hydraulically extended and mechanically retracted by redundant springs. A retractable arrester hook, for carrier landing operations, is mounted under the empennage, between the engine nacelles. This consists of a retract actuator/damper, a fail safe manual latch and release, a universal hook shank pivot, and a replacement hook point. Hook control is a manual system which automatically extends the hook in case of a release system failure.

Miscellaneous Systems: The General Electric fly-by-wire flight control system is of the digitized quadruplex type offering quadruple redundancy through a digitized multiplex bus. It has a direct electrical back-up to all control surfaces and a direct mechanical back-up to the slab stabilators. Through the quadruplex system, controls are optimized to each flight condition automatically.

There are two pitot static tubes mounted under the aircraft nose, on each side forward of the nose wheel well. Each tube contains one pitot source and two static sources.

The fire detection and extinguisher system is made up of three fire warning/extinguisher lights, a fire extinguisher pushbutton, one fire extinguisher bottle, a fire test switch, and dual-loop fire detection sensors. The extinguisher bottle is in the aft fuselage between the engines. The bottle contains a nontoxic gaseous agent which provides a one-shot extinguishing capability.

Air conditioning is provided by a Garrett unit and electrical power is provided by two General Electric VSCF 40 kVA generators. DC electrical power in aircraft BuNo. 161353 through BuNo. 161528 is provided by two transformer-rectifiers and two batteries with integral battery chargers; aircraft BuNo. 161702 and up have two transformer-rectifiers and two 7.5 Ampere-hour sealed lead acid batteries with a single battery charger.

Exterior lighting consists of position lights, formation lights, strobe lights, an arresting hook floodlight, and a

F/A-18A of VFA-151 shortly after becoming airborne. Leading and trailing edge flap and flaperon extension angles is noteworthy, as is squadron badge art visible on wing tank. Nose gear strut is seen at maximum extension.

F/A-18A, BuNo. 162891 of VFA-161 at NAS Dallas on August 23, 1986. Lightning bolt and tail code are dark gray against medium gray background. All other markings are standard for type, including "101" on nose and outboard section of each flap.

F/A-18A, BuNo. 162878 of VFA-192 at NAS Fallon during October 1986. This aircraft is equipped with a single AGM-88A "HARM" (High-Speed Anti-Radiation Missile) on its port outboard wing pylon. The weapon is optimized to destroy radar sites.

F/A-18A, BuNo. 162828 of VFA-195 during a transient stop at NAF Andrews on July 12, 1986. Placement of the squadron tail coding between the vertical formation strip light and the rudder leading edge is noteworthy.

F/A-18A, BuNo. 162837 of VFA-195. Markings are dark gray on gray and thus typical for type. Aircraft is being flown clean, except for centerline drop tank. Nose cap and centerline tank nose cap are beige.

F/A-18A, BuNo. 161735 of VFA-303. Folded outer wing panel gives some indication as to the scarcity of "Hornet" undersurface markings. Markings are standard dark gray on gray with the exception of the black "301" and "01".

F/A-18A, BuNo. 161716 of VFA-303 on January 30, 1987. Early style markings are apparent, with very light gray to white undersurfaces and wing tanks, and medium gray upper surfaces. Miscellaneous identifiers and squadron badge are in dark gray.

F/A-18A, BuNo. 161712 of VFA-305. Bearing early gray on gray camouflage, aircraft is carrying a multiple ejector rack under each outboard wing pylon and a single Kelvin ACMI (Air Combat Maneuvering Instrumentation) pod on its port wingtip.

refueling probe light; interior lighting consists of console lighting, instrument lighting, flood lighting, utility flood lighting, engine instrument lighting, warning/caution lighting, and emergency instrument lighting.

Hydraulic power is supplied by two separate 3,000 psi systems capable of producing a maximum flow rate of 36 gpm. Each system consists of two hydraulic circuits. The two hydraulic systems are identical with the exception of the fluid supply line from the hydraulic system 2 reservoir assembly to the APU hydraulic hand pump (furnishing fluid at a pressure of 85 psi). The left, or system 1, provides power to the primary flight control surface actuators exclusively. The right, or system 2, also provides power to the primary flight control actuators but additionally supplies power to the speed brake and non-flight control actuators. Redundancy to the flight control actuators is achieved either by simultaneously pressurizing the actuator from both systems or by supplying pressure to the actuator from one system while the other system is in a back-up mode.

The 350 lb./4.4 cu. ft. Hughes aircraft AN/APG-65 coherent pulse-Doppler multi-mode digital radar has air-to-ground and air-to-air modes. The air-to-air mode, which uses interleaved high and medium-pulse repetition frequency for all-aspect target acquisition and pulse-Doppler for AIM-7F requirements, includes velocity search (VS), range while search (RWS), track while scan (TWS—which permits the tracking of ten targets while displaying eight), and a raid assessment mode (RAID) to break out targets flying in formation at greater range. It is fully digital in operation with its digital converter being part of the receiver/exciter unit, and it has a programmable signal processor (for the first time in a production radar) allowing relatively quick and easy software changes to be made if the threat or performance of new radar-guided weapons demands such.

There are three pilot-selectable air combat maneuvering (ACM) modes in which the radar automatically locks on and tracks airborne targets. The ACM modes are Wide Acquisition (WACQ), Vertical Acquisition (VACQ) and Boresight (BST). The ACM modes enable the pilot to be in head-up operation while target acquisition and launch envelopes for the selected weapons are displayed on the HUD. The ACM modes are weapon dependent. A fourth ACM mode is initialized when the pilot selects the gun as a weapon and is not in single target track. In this mode, the radar searches the HUD field of view in azimuth and elevation and 10 nm in range. The radar provides range, range rate, angle, and angle rate data to facilitate a high stabilized gun solution with minimum settling time, thereby providing an excellent director gunsight.

A fifth auto acquisition is provided the pilot to minimize his workload. This mode is selected from the sensor control switch which is located on the flight control stick. This mode is the Auto Acquisition mode (AACQ) and is selectable when the radar is operating in Range While Search (RWS) or Track While Scan (TWS) modes. When a mode is selected, the radar is commanded to acquire and track the first target it detects.

The air-to-surface modes include a real beam ground mapping mode, with three additional levels of "zoom" capability, two via Doppler Beam Sharpening (DBS) at ratios of 19 to 1 and 67 to 1 out to 40 nm, plus a Synthetic Aperture Radar Mode (SAR) giving 30 ft. by 60 ft. resolution out to 30 nm. These provide excellent cockpit imagery. Additional features, linked with the weapons/fire control system, are air-to-ground ranging, fixed target track, and precision velocity update. Sea surface search/ track, ground moving target identification/track, as well as Position Velocity Update (PVU) are all available.

The aircraft is equipped with an automatic carrier landing system (ACLS) for all-weather carrier operations; an Itek AN/ALR-67 radar warning receiver system; the ITT/Westinghouse AN/ALQ-165 airborne self-protection jammer (ASPJ) system; two AN/AYK-14 digital computers; a Litton AN/ASN-130A inertial navigation system; two Kaiser multi-function CRTs with a central Ferranti/ Bendix CRT and head-up display; a Conrac communications control system; a Normalair-Garrett digital data recorder for the Bendix maintenance recording system; a flight incident recording and monitoring system (FIRAMS); a Smiths standby altimeter and Kearflex standby airspeed indicator, standby vertical speed indicator, and cockpit pressure altimeter; conventional flight instruments and navigation equipment including a TACAN, an AN/APN-194 radio altimeter, two UHF radios, a UHF data link, and an IFF system; and a Garrett gas turbine APU (weighing 112 lbs. and generating 200 hp) for engine starting and ground pneumatic, electrical, and hydraulic power.

As a point of interest, it should be mentioned the F/A-18 has 307 access doors. Over 90% of these can be reached from ground level without work stands.

Armament: The F/A-18C is equipped with nine external weapons stations with a combined capacity of 17,000 lbs. This weight, which can consist of mixed ordnance, can be carried throughout a significant proportion of the aircraft's high-g flight envelope. Available stations include the wingtips (for AIM-9/L/M only), the two outboard wing pylons (available for AIM-7F/M, AIM-9/L/M, AIM-120, and AGM-65E/F and air-to-surface weapons); two inboard wing pylons for external fuel tanks or a variety of air-to-surface weapons (including AGM-88A HARM and AGM-84 *Harpoon*), two nacelle fuselage stations (for either two AIM-7Fs or on the right station a Hughes

F/A-18A, BuNo. 161248 of VX-4. Early light gray to white on gray markings are apparent. Tail code, insigne, and other miscellaneous markings are in dark gray. Noteworthy is VX-5 Playboy bunny on lower third of rudder.

F/A-18A, BuNo. 161248 of VX-4 as it departs Offutt AFB following a transient stopover. Aircraft mounts a single 330 gal. drop tank under each wing and a single AIM-9L "Sidewinder" on each wingtip rail.

F/A-18A of VX-5 at flare during early carrier trials. The two outboard pylons and the centerline pylon each mount horizontal ejector racks. Additionally, the wing drop tanks are of the original, oval type—a design later rejected.

F/A-18A, BuNo. 161736 of the NSWC at NAS Fallon on August 16, 1986. Markings are dark gray on medium gray, with a very light gray to white undersurface. The centerline tank is medium gray with a beige nose cap.

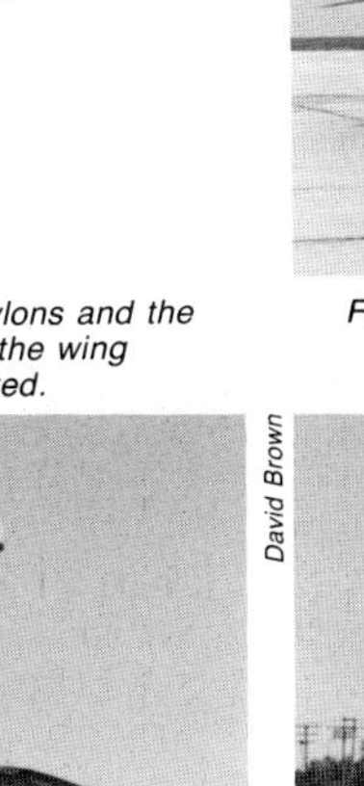

F/A-18A, BuNo. 161728 of the Navy Strike Warfare Center transient at Luke AFB on February 23, 1986. De-energized flap and flaperon droop is noteworthy. Lightning bolt and ''strike'' are in dark gray with a medium gray background.

F/A-18A, BuNo. 160782 (#7 F/A-18A proto.) of the NATC Strike Test unit at Patuxent River during October 1986. Noteworthy are the over-all medium gray markings and the orange vertical fin flashes. Cameras are attached to wingtip mounts.

AN/AAR-50 TINS pod, or a Martin Marietta AN/ASQ-173 laser spot tracker/strike camera pod (enabling cooperative precision bombing with laser designated targets/providing damage assessment imagery recorded for postflight review) and on the left station a Ford AN/AAS-38 FLIR pod (providing imagery displayed in the cockpit during night and low visibility conditions; it is capable of several sweep and track modes, including automatic target tracking with both a wide [12°] and a narrow [3°] field of view, it also has a coaxial mounted laser designator—this allows self-designation for laser guided weapons; and a centerline fuselage station for external fuel, a single nuclear store (B-57 or B-61/BDU-12 or BDU-20 or BDU-36 practice weapons), or various optional weapons. Miscellaneous weapons that have been cleared for use with the F/A-18 include the GBU-10 LGB, the GBU-23 LGB, the Mk 82 LGB, the Mk 82 GPB (low drag or high drag), the Mk 82 Retard GPB, the Mk 83 LGB, the Mk 83 GPB (low drag), the Mk 84 LGB, the Mk 84 GPB (low drag), the AGM-62 *Walleye* (and data link pod), flare dispensers, Mk 76 and Mk 106 dispensers (practice bombs), miscellaneous rocket pods (LAU-10D/A, LAU-61A/A, or LAU-68B/A), CBU-59/B *Rockeye*, BLU-95 FAE II (fuel/air explosive), and the CBU-59.

The F/A-18 also carries a General Electric M61A1 20mm rotary cannon with 570 rounds in a nose compartment. It is capable of firing at rates of up to 6,000 rpm. Aiming is accomplished via a McDonnell Douglas director gunsight; a conventional optical sight is available as a back-up.

POWERPLANT:

The F/A-18A/B/C/D is powered by two General Electric F404-GE-400 low-bypass afterburning turbofan engines each capable of generating 10,700 lbs. thrust dry, and in excess of 16,500 lbs. thrust in full afterburner. These engines have several remarkable performance characteristics, not the least of which is their ability to go from idle to full thrust in some four seconds. In combat, with low fuel and modest armament, the F404 gives the F/A-18 a thrust-to-weight ratio of greater than one-to-one.

Evolved directly from the AF YJ101 developed for the Northrop YF-17, the F404 offers a major improvement in bypass ratio (0.20 to 0.34) while requiring only a 1 in. increase in fan diameter and a slightly larger low pressure turbine section.

Because of the unique environment generated by carrier operations, Navy-qualified engines are put through a rigorous test program to certify their dependability. The F404 eventually logged no less than 14,000 factory test hours utilizing 14 development engines over a period of a little less than five years. The ground test program was initiated during December 1976 and ten engines had been completed for test work by the first few months of the following year. By that time, some 3,500 test hours had been accumulated, including 1,500 hours of endurance testing. Preliminary flight rating tests took place during May 1978, and the first flight powering an F/A-18 followed during November. Nine engines were delivered to MCAIR during 1978, and another 24 followed during 1979. Model qualification tests were completed during July 1979 with the first production engines delivered to MCAIR during December.

Air is supplied the engines through a pair of semi-elliptical fixed-ramp intakes located one beneath the midsection of each LEX. These are designed to provide compatible air to the engine. The system uses a fixed geometry compression ramp, a fuselage boundary layer diverter system, and a ramp boundary layer bleed system. The compression ramp provides the correct oblique shock up to the maximum aircraft design speed of Mach 2.1. The fuselage boundary layer diverter system prevents low energy air from entering the inlets. This air is diverted below the fuselage. The rear part of the compression ramp is porous to prevent this boundary layer air from entering the inlet. Part of the boundary layer air is bled through a fixed area outlet into the fuselage boundary layer diverter channel. The other part exits on top of the wing through inlet duct doors, when open. These automatically open at Mach 1.33 (accelerating) and close at Mach 1.23 (decelerating). The doors are controlled by the flight control computer.

The engines are equipped with an airframe-mounted accessory drive (AMAD) which powers the fuel pumps, hydraulic pumps, and generators during flight; on the ground it can be used to test various aircraft functions without having to actually start the engines. Also, engine starting ''yellow gear'' is eliminated with the use of an auxiliary power unit. Under ideal conditions, an engine can be changed in less than 15 minutes.

In the F/A-18C, the two ejector pumps found on the earlier F/A-18A have been replaced with turbine transfer pumps that allow for better fuel utilization and distribution. These totally independent pumps maintain a balance of fuel in each of the F/A-18C's four internal tanks, thus eliminating imbalances that might cause problems during air combat.

The F/A-18 is equipped with self-sealing fuel tanks and fuel lines with fire suppression foam in the wing tanks and fuselage void areas. Internal fuel capacity is 1,589 gals. (10,330 lbs. JP-4 or 10,810 lbs. JP-5) for the FA-18A/C (the F/A-18B/D carry only 316 gals. in tank #1) and up to three 315 gal. (elliptical) or three 330 gal. (cylindrical) or three 480 gal. (Australian use) external tanks can be carried as well (with the 330 gal. tanks, the total capacity is 17,800 lbs.). The fuel is carried internally in four interconnected fuselage tanks and two internal wing (wet) tanks. External tanks may be mounted on the centerline and inboard wing station pylons.

A single refueling point on the left side of the front fuselage is used for both ground refueling and fuel system purging. Fuel dumps and vents are located on the top of the vertical tail surfaces. The aircraft also is equipped with a retractable, inflight refueling boom on the right side of the nose. The fuel quantity monitoring system is by Simmonds.

Advanced F404 configurations for use with advanced F/A-18s currently are in development. Maximum afterburning thrust ratings of these proposed engines (scheduled for availability by the year 2,000), with engine mass flows approaching 200 lbs. per sec., increased combustor pressure, improved high pressure and low pressure turbines, increase turbine inlet temperatures, and improved hot section materials now is estimated to be in the 23,000 lb. range. Additionally, near-term modifications to the F/A-18 to solve a chronic engine ice ingestion problem, in the form of an electro-expulsive deicer under development by the NASA, are presently being considered.

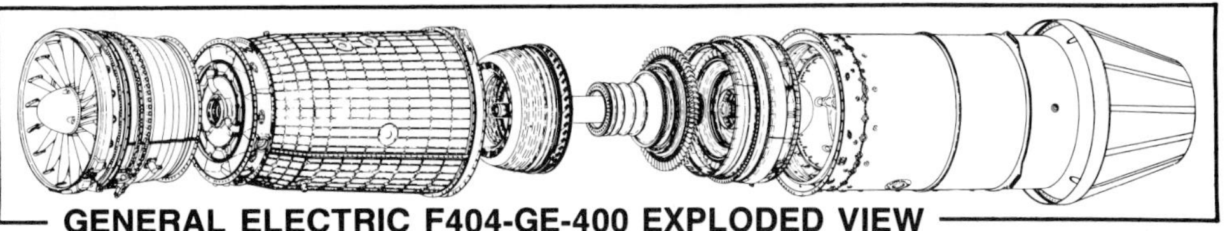

F/A-18A, BuNo. 161702 of the NATC Strike Aircraft Directorate at NAS Patuxent River on September 2, 1985. The aircraft is carrying a pair of Mk.82 "Snakeye" retarded bombs on its starboard outer wing pylon and an unidentified weapon inboard.   .

F/A-18A, BuNo. 161925 of the NATC Strike Aircraft Directorate certainly is the most colorful "Hornet" yet. Nicknamed "D'Skunk", it is black and white with each quadrant of the aircraft, when viewed head-on, being an opposing color.

F/A-18A, BuNo. 161713 of the Naval Weapons Center at China Lake transient at NAF Andrews. NWC badge barely is visible on the vertical fin. All other markings, though of early vintage, are standard for type.

F/A-18A, BuNo. 161366 of the Naval Weapons Center at China Lake. Aircraft is carrying a dummy AIM-9L on its wingtip rails, a single Mk.84 "Paveway" 2,000 lb. laser guided bomb, and an AGM-88A "HARM" on its port inboard wing pylon.

A Pacific Missile Test Center F/A-18A set up for an airshow display with a wingtip rail mounted AIM-9L, a pylon mounted AIM-120 AMRAAM, a pylon mounted AIM-7E "Sparrow", and a pylon mounted AGM-88A "HARM" under the port wing.

The ninth F/A-18A FSD aircraft, BuNo. 160785, at NASA's Dryden facility at Edwards AFB. This is one of several F/A-18s scheduled for utilization by NASA as replacements for their rapidly aging fleet of Lockheed F-104 "Starfighters".

The prototype F/A-18A, BuNo. 160775, which also was the first of the FSD aircraft, at NASA's Dryden/Edwards AFB facility. This "Hornet" will be utilized as a spare parts aircraft for the rest of the NASA "Hornet" inventory.

The fourth F/A-18B, BuNo. 161249, currently is being flown by the Naval Test Pilot School at NATC Patuxent River. The basic markings are red and white, with black trim lines and a black anti-glare shield.

Front view of F/A-18B, BuNo. 161249 illustrates flap and flaperon marking details not discernible in side view. Trim is bright red, with black trim lines delineating red areas from white. Yaw string angle marks are discernible on anti-glare panel.

"Blue Angels" #4 during 1987 Carswell AFB airshow. Markings are bright yellow against an over-ali gloss Navy blue. "Blue Angels" aircraft entail several modifications not found on standard operational "Hornets".

F/A-18A ''Blue Angels'' #4, is not necessarily the same aircraft seen in the preceding photo. This #4 is BuNo. 161524. ''Blue Angels'' tail identification numbers periodically are switched from one aircraft to another as support requirements dictate.

CF-18A, s/n 188731, of 410 Sq. Canadian Armed Forces during June 1986 at Cambrai, France. Canadian ''Hornets'' are camouflaged dark gray against a medium gray. The nose cone cap remains beige. ACMI pod on starboard wingtip is noteworthy.

CAF CF-18B, s/n 188911 during 1987 William Tell meet at Tyndall AFB. The Canadian Air Force historically has participated repeatedly in this important competition and has maintained an excellent scoring record.

CAF CF-18A, s/n 188718 of No. 425 Sqdn. Visible near the tip of the vertical fin is No. 425 Sqdn. bird logo. Use of independent crew boarding ladders for ingress and egress during 1987 William Tell meet is noteworthy.

RAAF F/A-18A, A21-8, of No. 3 Sqdn. This was one of the first two ''Hornets'' delivered to No. 3 Sqdn. arriving during the autumn of 1986. Australian aircraft have their insignia on the undersides of the wings, only.

CAF CF-18A, s/n 188767 during a transient stopover at Offutt AFB. Aircraft is clean except for centerline drop tank. Readily visible is undersurface camouflage pattern near nose gear designed to visually replicate cockpit/canopy.

RAAF F/A-18B, 21-105 of No. 2 OCU. Unit insigne on outer vertical fin surfaces is black and yellow. Serial numbers are black and the national markings are in full color. Australian camouflage is significantly more contrasty than U.S. Navy.

Spanish AF C.15, s/n C.15-13 of Escuadron 122/12 Wing at Torrejon shortly before delivery from MCAIR. The light gray over-all camouflage is standard for type in SAF service. All markings, excepting the national insigne, are in black and gray.

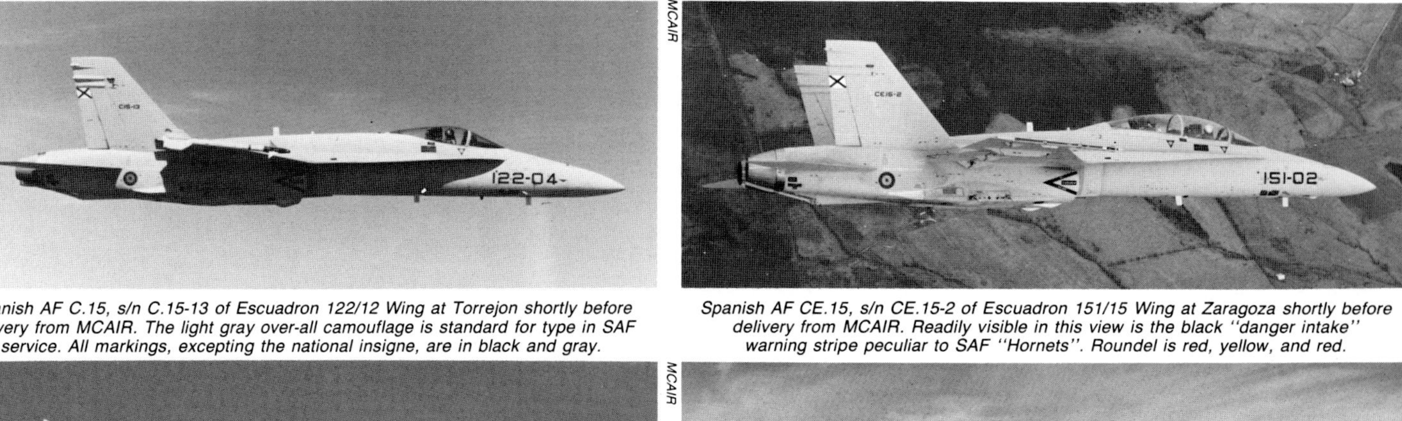

Spanish AF CE.15, s/n CE.15-2 of Escuadron 151/15 Wing at Zaragoza shortly before delivery from MCAIR. Readily visible in this view is the black "danger intake" warning stripe peculiar to SAF "Hornets". Roundel is red, yellow, and red.

Spanish AF CE.15s, s/ns CE.15-3 and CE.15-4 of Escuadron 151/15 Wing at Zaragoza during company photo session and prior to delivery from MCAIR. Squadron assignment is prominently displayed in black on noses of all SAF "Hornets".

The first "Hornet" to be equipped with the nose-mounted reconnaissance system was F/A-18A BuNo. 161214. It followed in the footsteps of the F/A-18A prototype, BuNo. 160775, which after modification had served as the initial aerodynamic prototype.

F/A-18A, BuNo. 161214 became the prototype F/A-18(R). This aircraft served to explore the production airframe changes required for reconnaissance capability to be added to the later F/A-18C/D series "Hornets".

The first production F/A-18C, departing MCAIR on its first flight, which took place on September 3, 1987. Externally visible differences between the F/A-18C and the F/A-18A are discernible in the form of additional EW system antenna fairings.

The first fully-equipped, night-attack capable F/A-18D, BuNo. 163434, departing St. Louis on its initial test hop on May 6, 1988. A Hughes AN/AAR-50 TINS pod is suspended from the starboard nacelle.

The second **YF-17, 75-1570**, in over-all Navy-demo gray scheme at NAS Lemoore, California on May 17, 1978. This aircraft effectively was utilized as the F/A-18 prototype and served to indoctrinate Navy pilots on the basic flight characteristics of the production aircraft. Though smaller and lighter, its performance envelope and flight control system feel were quite similar, and thus suitable for the indoctrination job. This aircraft presently is on display at the Naval Aviation Museum, Pensacola, Florida.

The prototype **F/A-18A, BuNo. 160775**, at Richards Gabaur AFB, Missouri on August 6, 1983. It is not well known that this aircraft was utilized as the aerodynamic prototype for the photo-reconnaissance ''Hornet'' and was equipped with the tell-tale bulged nose sensor compartment and associated transparencies. It now is part of the NASA ''Hornet'' inventory and is awaiting restoration to flightworthy status at NASA's Dryden facility at Edwards AFB, California.

F/A-18A, BuNo. 161703, assigned to the Pacific Missile Test Range with a Beechcraft BQM-126A target drone mounted under its starboard wing. The BQM-126A is replacing the Teledyne Ryan BQM-34 ''Firebee''. It has a 10 ft. wingspan, is 18 ft. long, weighs 1,398 lbs. with booster, and has a range of 750 miles. At the time, the drone was being statically drop-tested utilizing a cushioned pit as the receptacle. The F/A-18A has an exceptional external load-carrying capability in excess of eight tons.

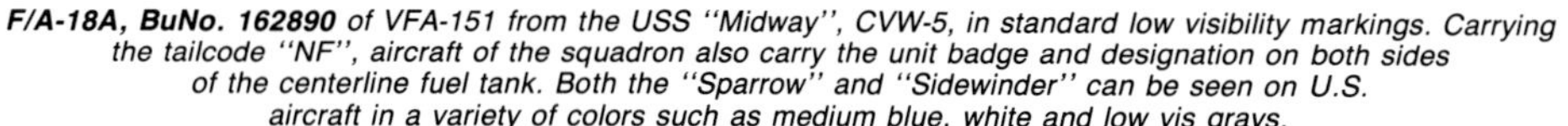

F/A-18A, BuNo. 162890 of VFA-151 from the USS "Midway", CVW-5, in standard low visibility markings. Carrying the tailcode "NF", aircraft of the squadron also carry the unit badge and designation on both sides of the centerline fuel tank. Both the "Sparrow" and "Sidewinder" can be seen on U.S. aircraft in a variety of colors such as medium blue, white and low vis grays.

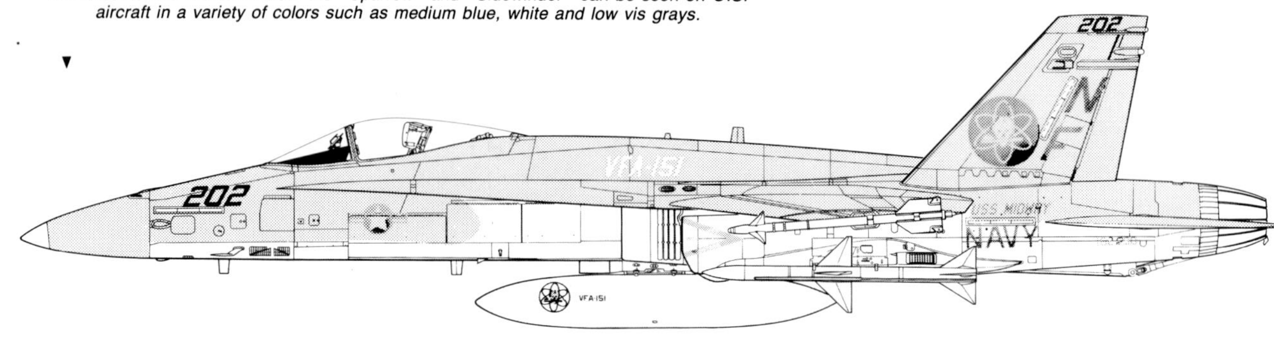

F/A-18B, BuNo. 161249 of VFA-125 in standard low visibility markings. Bearing the tail code "NJ", aircraft of VFA-125 carry Marines on their aft portside fuselage and Navy on their starboard aft fuselage in F.S. 36375 light compass ghost gray.

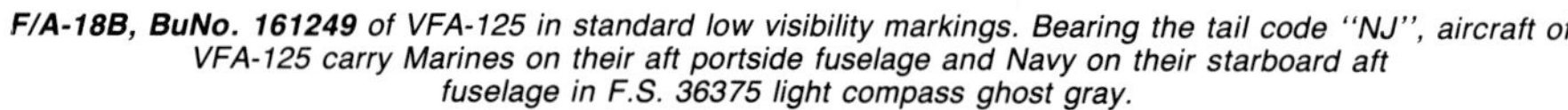

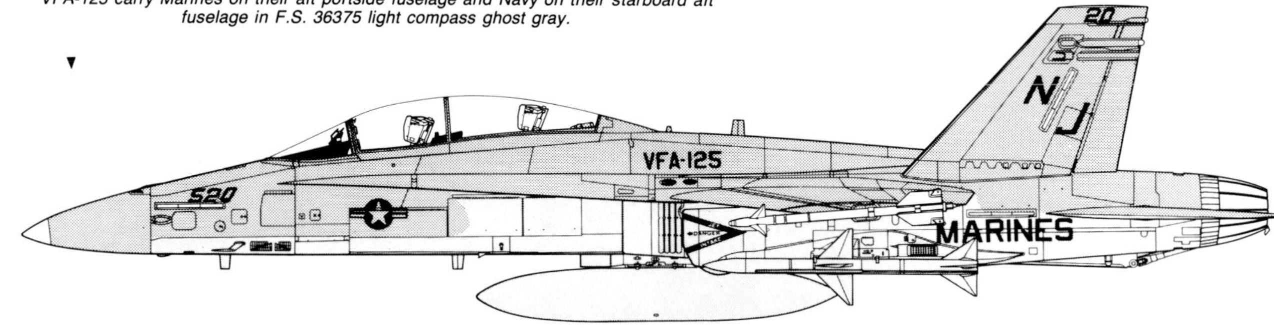

CF-18A, s/n 188712 of the Canadian "Forces Armees" No. 409 Squadron. Canadian low visibility paint schemes differ from those of U.S. aircraft and are distinguished by the under fuselage false canopy marking in F.S. 36118 sea gray. The false canopy markings are intended to momentarily disorient hostile pilots attempting to discern aircraft flight direction. Other aircraft colors are F.S. 35237 gray blue upper surfaces and F.S. 36375 light compass ghost gray undersurfaces.

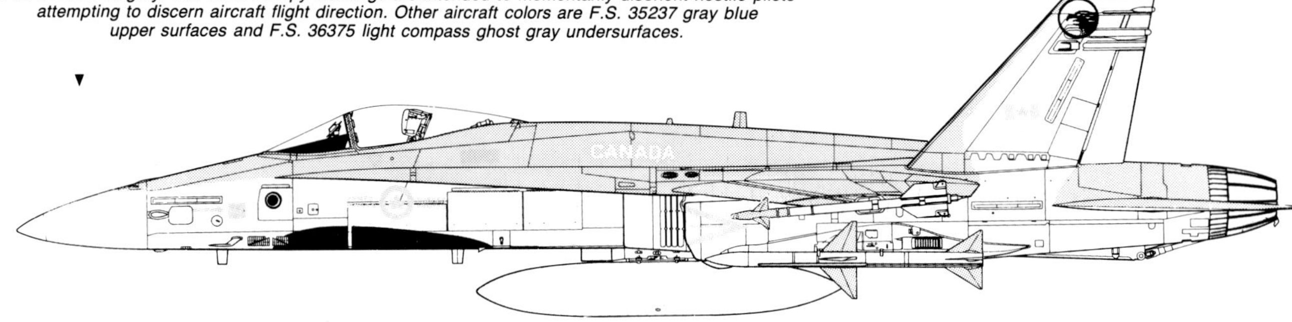

CF-18A, s/n 188713 of the Canadian "Forces Armees", Number 425 Squadron "Alouettes". Like all Canadian CF-18s, this aircraft has a searchlight installed in the forward portside fuselage ahead of and below the wing leading edge chine. The aircraft is in standard Canadian low visibility paint and carries the distinctive undersurface false canopy marking. Note that the leading edges of Canadian "Hornet" vertical fins are painted F.S. 35237 gray blue, while the rest of the fin surface is F.S. 36375 light compass ghost gray.

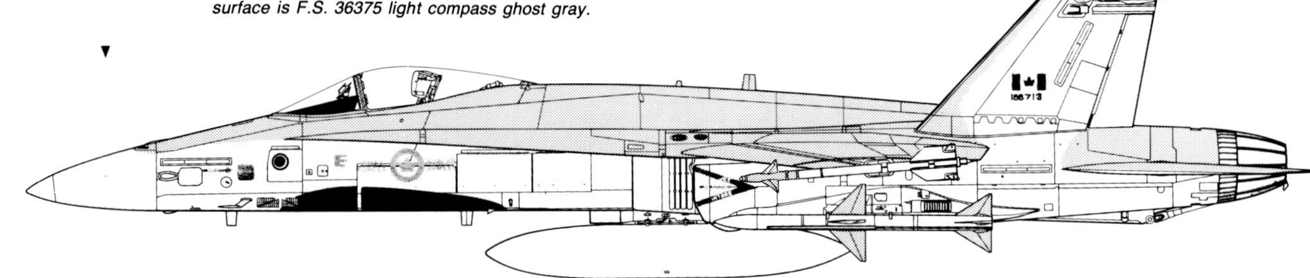

SELECT MARKINGS

Scale: 1/100th

Drawn by Mike Wagnon

F/A-18A, BuNo. 160776, the second F/A-18A FSD prototype in overall white with blue and gold flashes. Gold markings on this aircraft appear only on the leading edges of the horizontal stabilator while the balance of the stabilator is in blue and white, similar to the vertical fin. All other markings are standard for type.

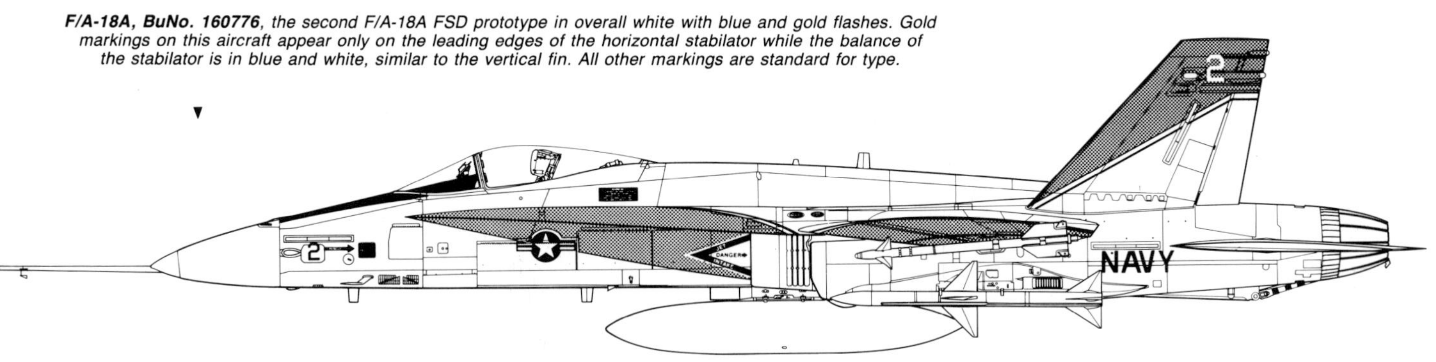

F/A-18A, BuNo. 161840 in overall white with blue flashes while assigned to the NASA for testing. Stylized NASA logo is in gloss black while the anit-glare panel appears to be light compass ghost gray. Aircraft data placards and national insignia are standard for type.

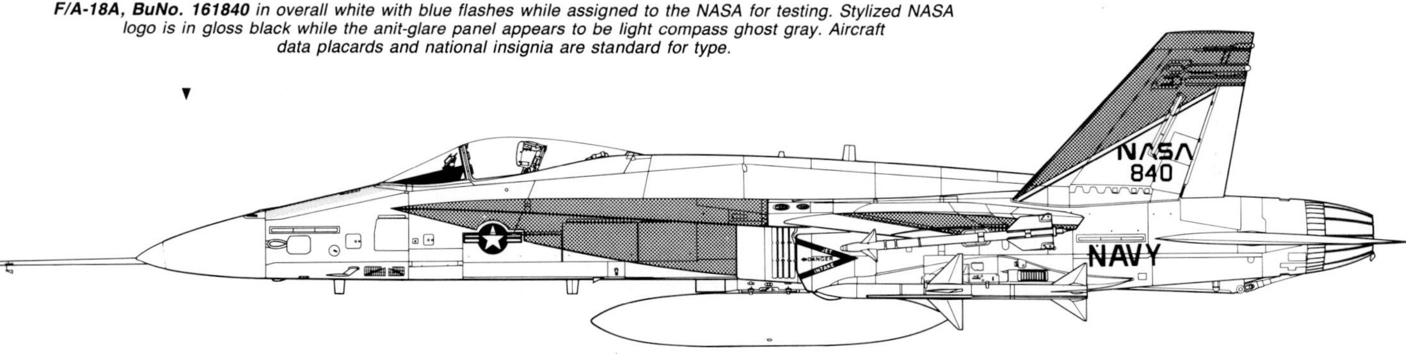

F/A-18A, BuNo. 161214 modified as the prototype photographic reconnaissance RF-18A. It appears in standard low visibility markings with F.S. 35237 gray blue radome and extended pitot boom. The aircraft carries both Navy and Marines markings respectively on its starboard and port aft fuselage.

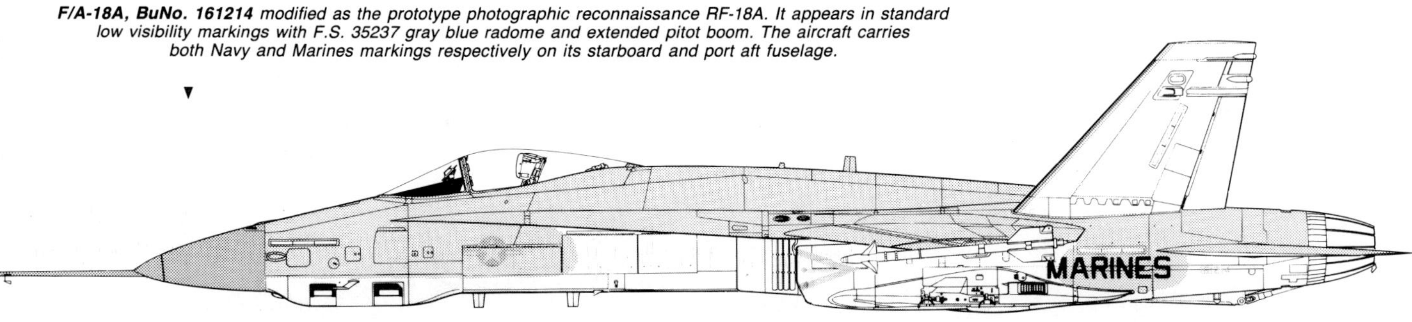

F/A-18A, BuNo. 161970 of VFA-131 ''Wildcats'' while aboard the USS ''Coral Sea''. Aircraft is in standard low visibility paint and carries the unit's distinctive feline insignia with the tail code ''AK'' in addition to the carrier name on the outside faces of both vertical fins.

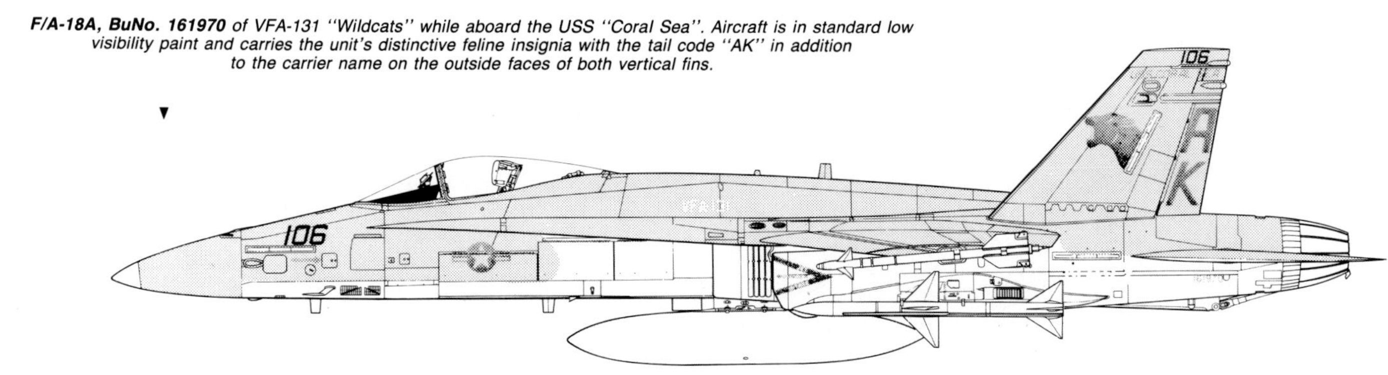

The Navy's ''Blue Angels'' aerobatic demonstration unit, during February 1986, formally committed to the acquisition and use of the ''Hornet'' as their team aircraft. Accordingly, an initial batch of seven high-time early production aircraft was allocated, these consisting of six F/A-18As and a single F/A-18B. Later, following the loss of one of the initial machines, the total contingent was increased to eight, with two aircraft being considered as spares.

Color, even in relatively small amounts, has been a rarity among Navy ''Hornets''. This brand new VFA-82 **F/A-18C, BuNo. 163456,** is one of the rare exceptions. New Navy regulations now permit two aircraft per fighter squadron to bear colorful markings—in moderation. Accordingly, color should become slightly more common during the coming years. The F/A-18C only recently has begun entering the operational Navy inventory. Noteworthy are this model's distinctive nose and fuselage antenna fairings.

F/A-18A, BuNo. 163092 of VFA-106. This aircraft is one of the rare ''Hornets'' to bear color in the form of tail art and more pronounced identifiers. A single TER (triple ejector rack) is visible under the starboard wing. TERs often are utilized to carry small training bombs as their shackle design is one of the few remaining in inventory with this capability. The ''Gladiators'' serve as the east coast training squadron for the ''Hornet'' and operate out of NAS Cecil Field, Florida.

F/A-18A, BuNo. 161925, of the NATC Strike Aircraft Directorate during May 1986. Nicknamed "D'Skunk" and tail-coded "7T", this remains perhaps the most gaudily painted "Hornet" presently extant. Basic markings consist of splitting the aircraft into four quadrants with each painted an opposing black or white. This theme continues onto the inside of each vertical tail surface but ends with the gray dielectric nose cone and the all-white centerline drop tank.

Temporary paint on F/A-18A, BuNo. 162395 was applied for combat exercises conducted from NAS Pt. Mugu during October 1985. Markings are limited to a water soluble light blue/aqua over the standard aircraft low visibility gray. Combatants are painted in these unusual schemes to differentiate them from enemy or friendly aircraft. Many variations to this theme have been seen over the past few years as the importance of practice maneuvers has grown in stature.

The first Royal Australian Air Force F/A-18B, A21-103, following acceptance ceremonies on May 4, 1985. Eighteen F/A-18Bs are scheduled for RAAF acquisition, along with 57 F/A-18As. Two of the F/A-18Bs were built by McDonnell Douglas at their St. Louis facility and delivered by air to Australia during a multi-stage trans-Pacific hop. These aircraft actually had been preceded into the air by the first indigenously-produced Australian "Hornets"—which were simultaneously absorbed into the RAAF inventory.

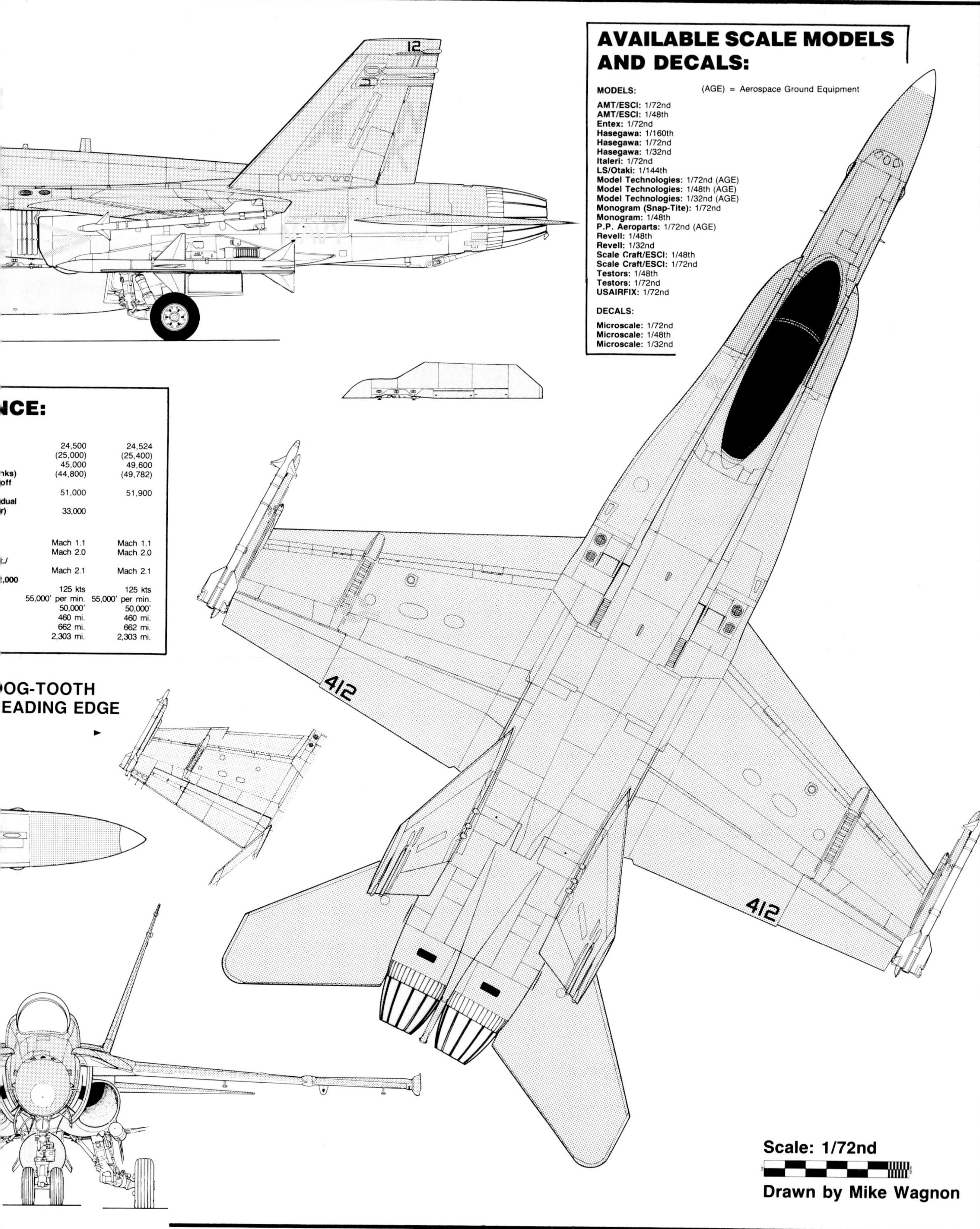
AVAILABLE SCALE MODELS AND DECALS:

MODELS:

(AGE) = Aerospace Ground Equipment

AMT/ESCI: 1/72nd
AMT/ESCI: 1/48th
Entex: 1/72nd
Hasegawa: 1/160th
Hasegawa: 1/72nd
Hasegawa: 1/32nd
Italeri: 1/72nd
LS/Otaki: 1/144th
Model Technologies: 1/72nd (AGE)
Model Technologies: 1/48th (AGE)
Model Technologies: 1/32nd (AGE)
Monogram (Snap-Tite): 1/72nd
Monogram: 1/48th
P.P. Aeroparts: 1/72nd (AGE)
Revell: 1/48th
Revell: 1/32nd
Scale Craft/ESCI: 1/48th
Scale Craft/ESCI: 1/72nd
Testors: 1/48th
Testors: 1/72nd
USAIRFIX: 1/72nd

DECALS:

Microscale: 1/72nd
Microscale: 1/48th
Microscale: 1/32nd

ICE:

24,500   24,524
(25,000)   (25,400)
45,000   49,600
(44,800)   (49,782)

51,000   51,900

33,000

Mach 1.1   Mach 1.1
Mach 2.0   Mach 2.0

Mach 2.1   Mach 2.1

125 kts   125 kts
55,000' per min.   55,000' per min.
50,000'   50,000'
460 mi.   460 mi.
662 mi.   662 mi.
2,303 mi.   2,303 mi.

DOG-TOOTH
LEADING EDGE

Scale: 1/72nd

Drawn by Mike Wagnon

McDONNELL DOUGLAS F/A-18A, BuNo. 161962

F/A-18A, BuNo. 161962 of VFA-25 "Fist of the Fleet" aboard the USS "Constellation". Low visibility colors are F.S. 35237, gray blue anti-glare panel; F.S. 36375, light compass ghost gray for top surfaces; and F.S. 36495, gray for lower surfaces and sides. Nose radome tip (represented in white) is F.S. 33613 radome tan. Decal markings are available on Microscale sheet number 72-550.

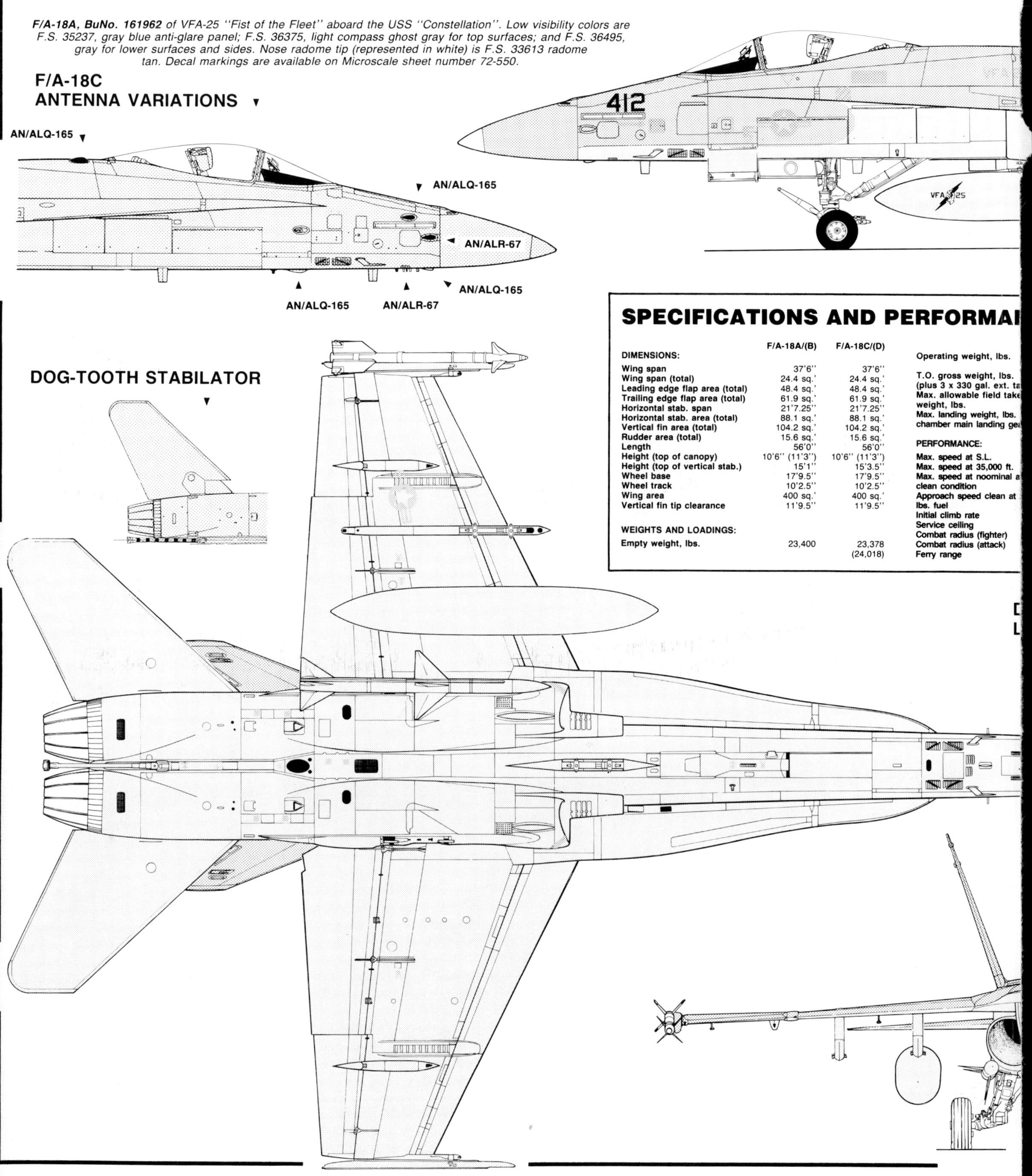

SPECIFICATIONS AND PERFORMA[NCE]

	F/A-18A/(B)	F/A-18C/(D)
DIMENSIONS:		
Wing span	37'6''	37'6''
Wing span (total)	24.4 sq.'	24.4 sq.'
Leading edge flap area (total)	48.4 sq.'	48.4 sq.'
Trailing edge flap area (total)	61.9 sq.'	61.9 sq.'
Horizontal stab. span	21'7.25''	21'7.25''
Horizontal stab. area (total)	88.1 sq.'	88.1 sq.'
Vertical fin area (total)	104.2 sq.'	104.2 sq.'
Rudder area (total)	15.6 sq.'	15.6 sq.'
Length	56'0''	56'0''
Height (top of canopy)	10'6'' (11'3'')	10'6'' (11'3'')
Height (top of vertical stab.)	15'1''	15'3.5''
Wheel base	17'9.5''	17'9.5''
Wheel track	10'2.5''	10'2.5''
Wing area	400 sq.'	400 sq.'
Vertical fin tip clearance	11'9.5''	11'9.5''
WEIGHTS AND LOADINGS:		
Empty weight, lbs.	23,400	23,378 (24,018)

Operating weight, lbs.

T.O. gross weight, lbs. (plus 3 x 330 gal. ext. ta[nks])
Max. allowable field take[off] weight, lbs.
Max. landing weight, lbs.
chamber main landing ge[ar]

PERFORMANCE:

Max. speed at S.L.
Max. speed at 35,000 ft.
Max. speed at noominal a[ltitude],
clean condition
Approach speed clean at [...]
lbs. fuel
Initial climb rate
Service ceiling
Combat radius (fighter)
Combat radius (attack)
Ferry range

*A colorfully marked **F/A-18A** of VMFA-312 carrying an unusual selection of weapons including wingtip-mounted AIM-9L ''Sidewinders'', a AGM-84 ''Harpoon'' under each starboard outer wing pylon, a single AGM-88 ''HARM'' under each port outer wing pylon, and two AIM-7F ''Sparrow'' AAMs on its fuselage fairings. This load underscores the diversity of weaponry capable of being carried by the ''Hornet''. Both the ''Harpoon'' and ''HARM'' are optimized as surface attack weapons.*

*The fourth full-scale-development (FSD) **F/A-18A**, BuNo. 160775 during its initial hop out of McDonnell Douglas' St. Louis, Missouri facility. The aircraft is completely unpainted with the exceptions of the national insigne, the number ''4'' on the vertical tail, and the black Navy bureau number just ahead of the slab stabilator. All natural metal paneling and composite paneling (dark colored) surfaces are visible. Noteworthy are the main landing gear in their fully extended, unloaded condition.*

*Cockpits of **F/A-18A**, BuNo. 162425 (left) and **F/A-18C**, BuNo. 163452 (right). The main panel configurations remain essentially similar, with three CRT-type primary displays occupying the center of the pilot's attention. Main difference lies in the engine and fuel quantity monitor panels which are visible just below the far left CRT. The F/A-18A's engine and fuel monitor is essentially analogue in display, whereas that for the F/A-18C is CRT/digitized.*

IN DETAIL:

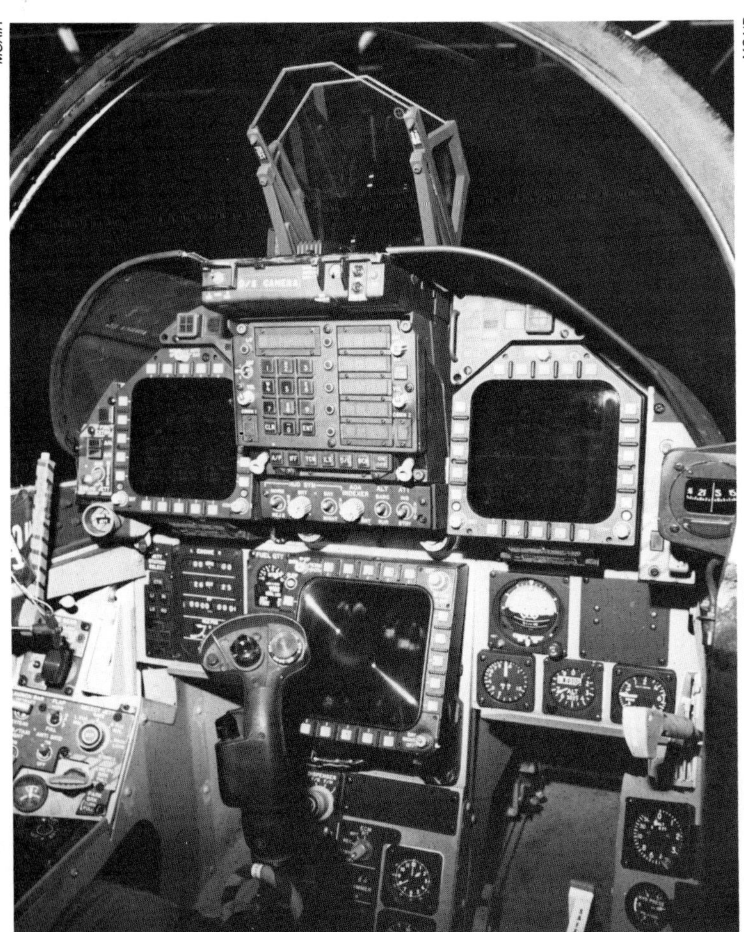

F/A-18A cockpit is dominated by three CRTs and their associated digitized presentations. Major flight and systems information is presented in digitized format on the CRTs with analogue instrumentation relegated primarily to emergency back-up.

Left F/A-18A console is mounting position for dual throttles which contain push-buttons and toggle switches which are integral with the Hands-On-Throttle-And-Stick (HOTAS) cockpit/pilot interface philosophy developed by MCAIR.

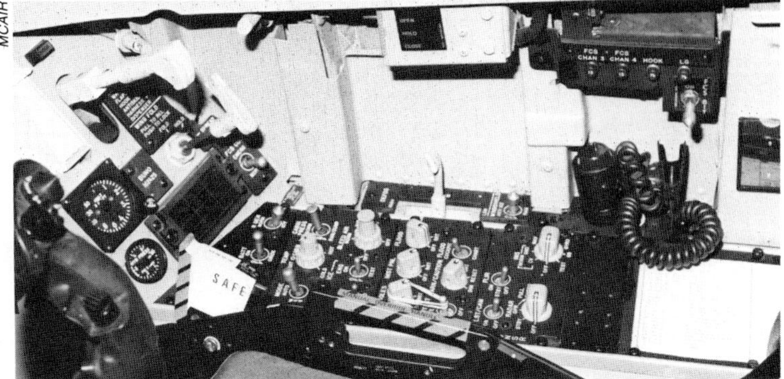

Right F/A-18A console is mounting position for LST/CAM, FLIR, and some radar switches, as well as INS switches and miscellaneous cockpit environmental controls. Large handle protruding from sub-panel on left controls tail hook.

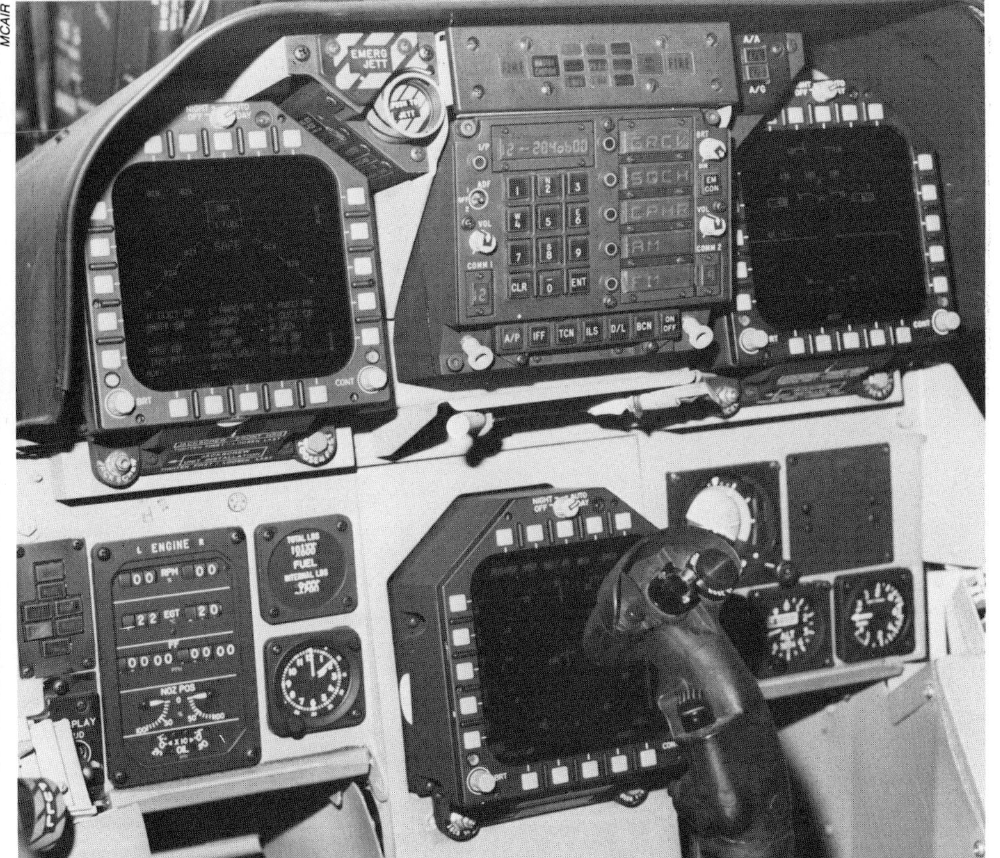

F/A-18B rear cockpit main panel. CRT layout and associated instrument spread essentially replicates that of front cockpit. Visible to the lower right are the analogue-style back-up attitude reference indicator, the standby altimeter, the standby airspeed indicator, and the standby vertical speed indicator.

F/A-18B rear cockpit left console, like that of the front cockpit, serves to support the dual throttles with their HOTAS buttons and toggle switches.

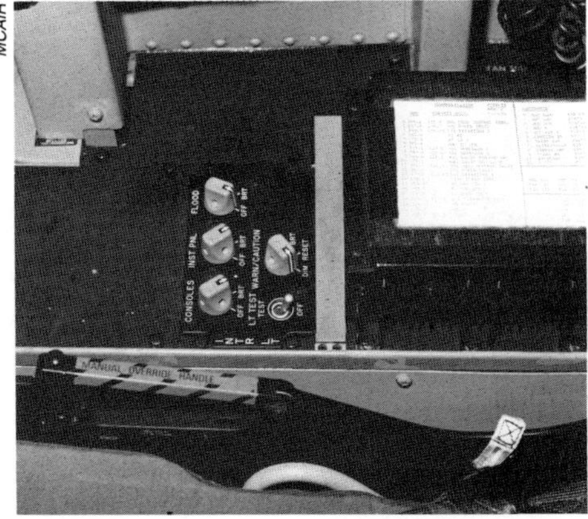

F/A-18B rear cockpit right console represents essentially unutilized space and mounts only a switch panel for interior lighting and a power supply unit.

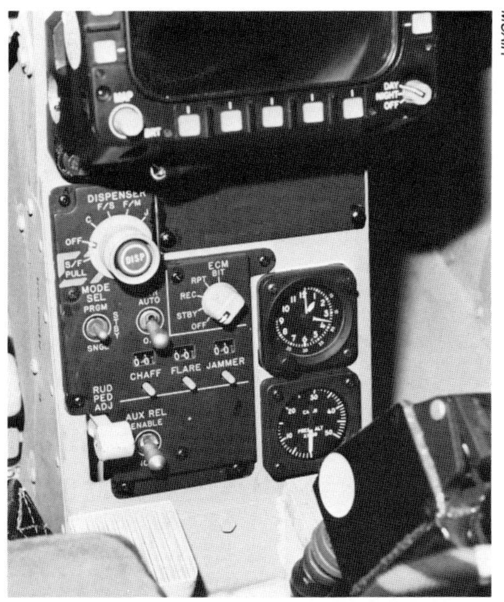

F/A-18A center sub-console, which fits between the pilot's legs, supports various defensive system controls, a clock, and a back-up altimeter.

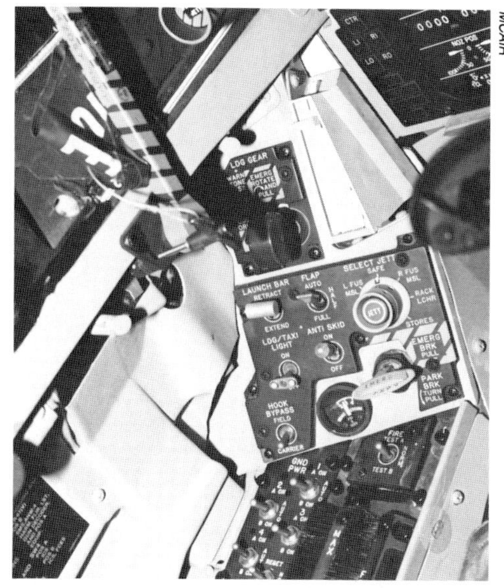

F/A-18A left sub main panel supports landing gear controls, the launch bar switch, the emergency/parking brake handle, and other miscellany.

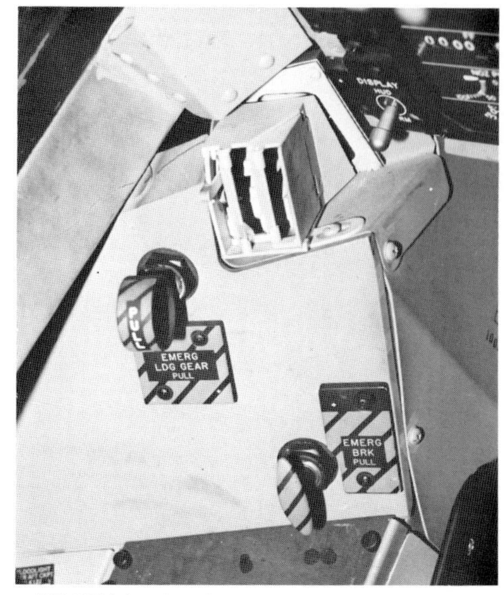

F/A-18B left sub main panel mounts ventilation vent, emergency landing gear brake actuation handle, and the emergency parking brake handle, only.

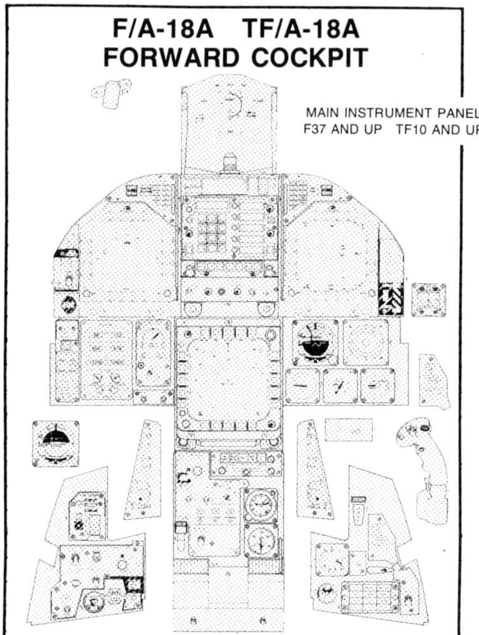

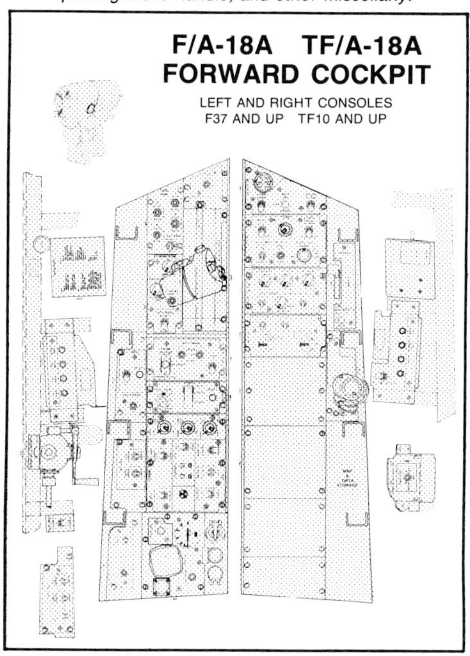

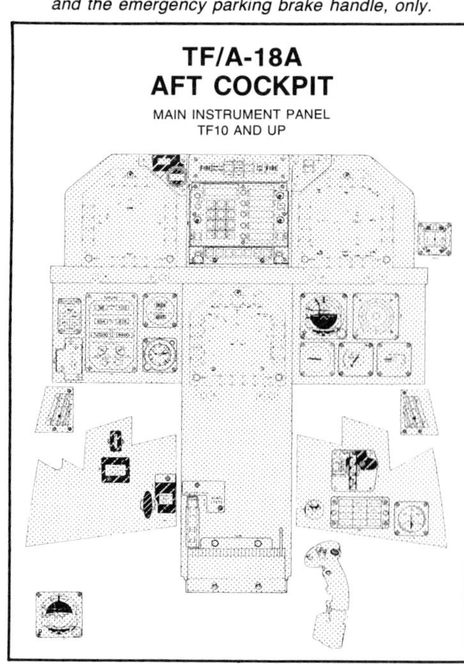

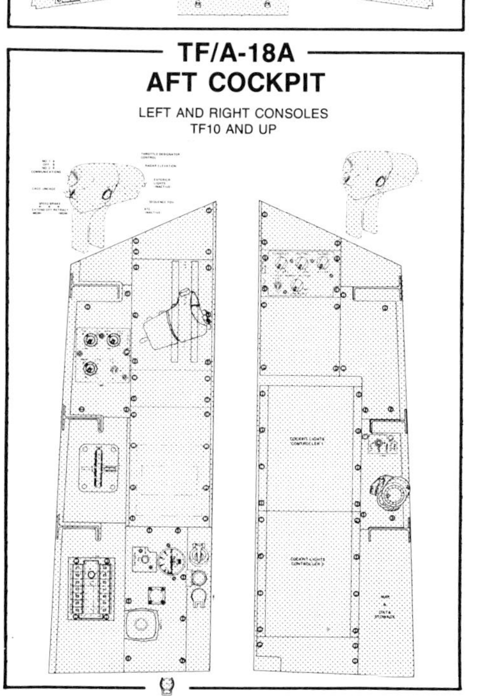

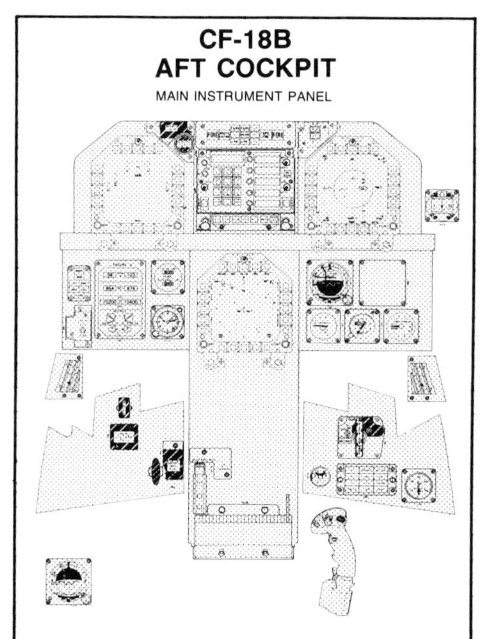

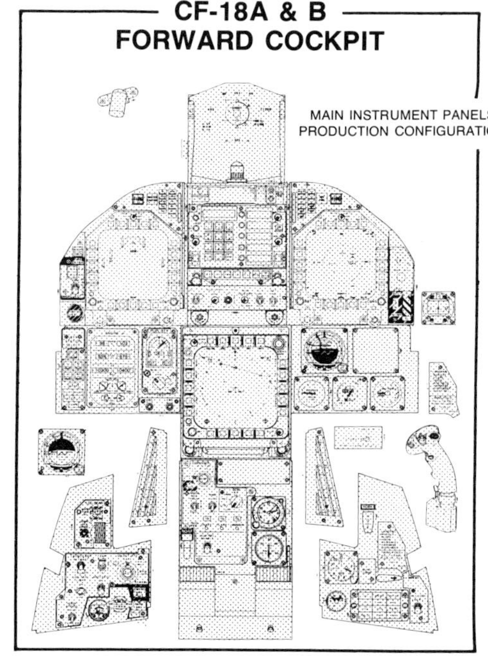

CF-18A & B FORWARD COCKPIT

CONSOLES
PRODUCTION CONFIGURATION

CF-18B AFT COCKPIT

LEFT AND RIGHT CONSOLES

MCAIR

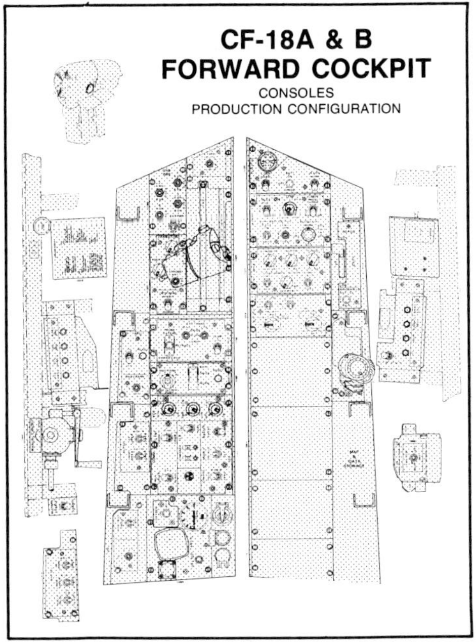

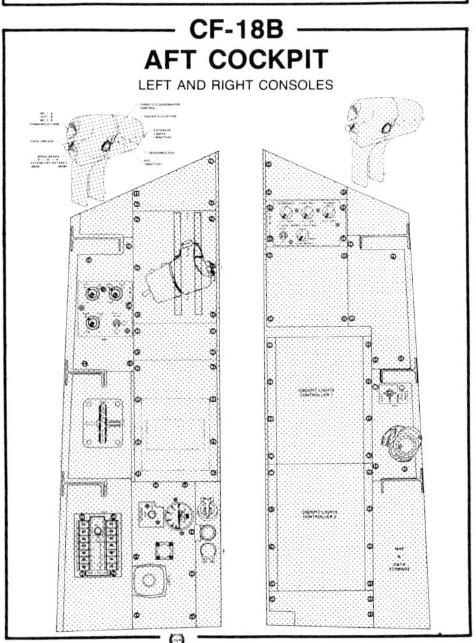

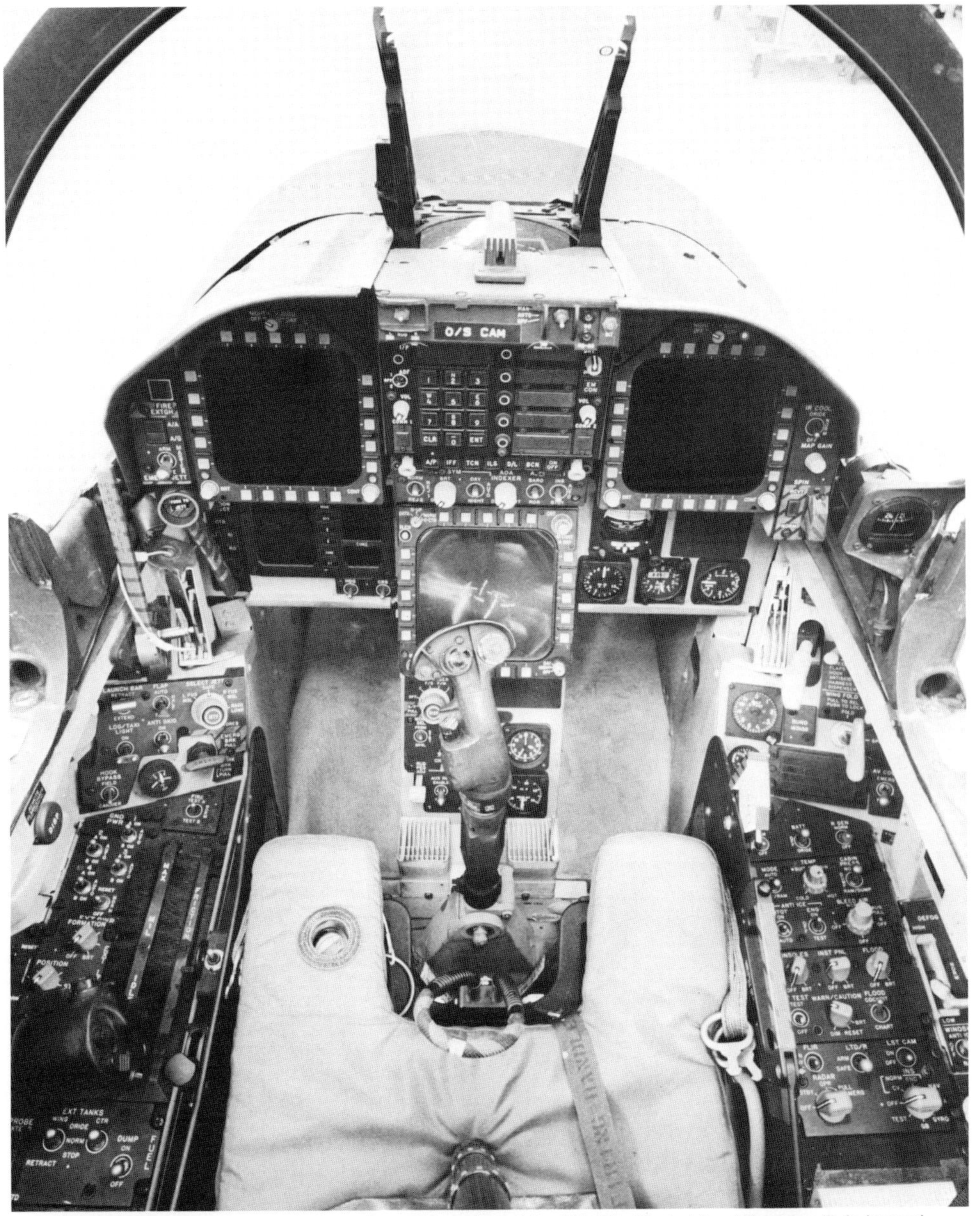

The F/A-18C cockpit essentially is unchanged from that of the F/A-18A. Most of the differences lie in internal hardware and software modifications, which are not readily visible. Perhaps the biggest physical change lies in the switch to a small CRT for the engine monitor display, located in the lower left segment of the panel.

MCAIR

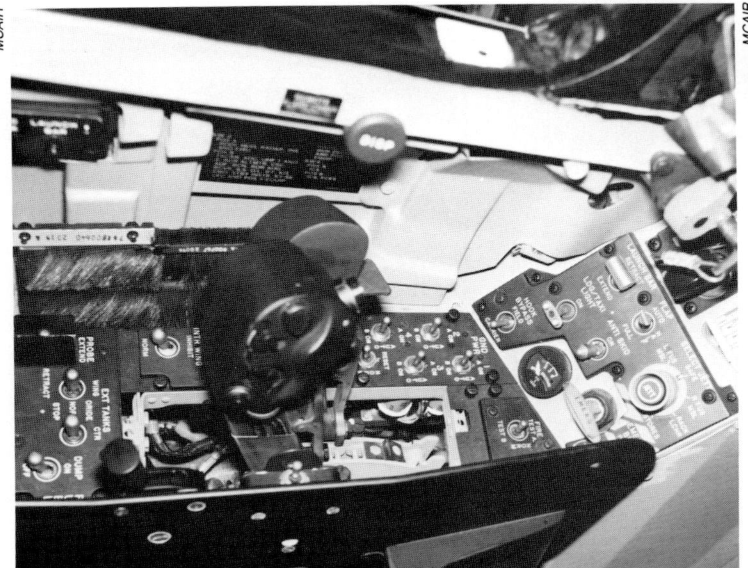

F/A-18C left console, minus cover plate, mounts HOTAS throttle with communications, speedbrake, sight, a/b finger lifts, chaff/flare dispenser, exterior lighting, APC, non-cooperative target/FLIR, and target designator control switches.

MCAIR

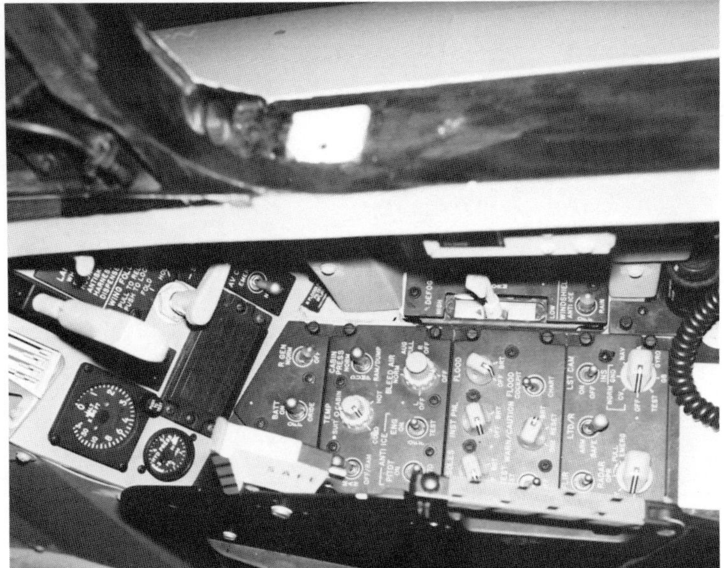

F/A-18C right console is little changed from F/A-18A and serves as mounting point for environmental control system panel, defog control, windshield anti-ice control, interior lighting panel, FLIR/LST panel, and other miscellany.

CREW STATION ORIENTATION FORWARD COCKPIT MAIN INSTRUMENT PANEL LEFT AND RIGHT VERTICAL CONSOLES

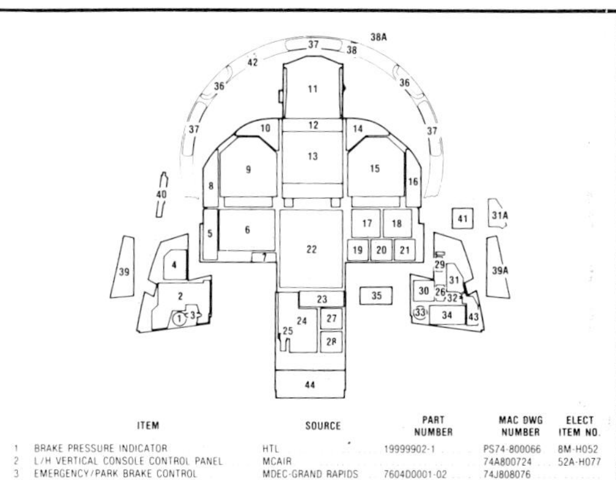

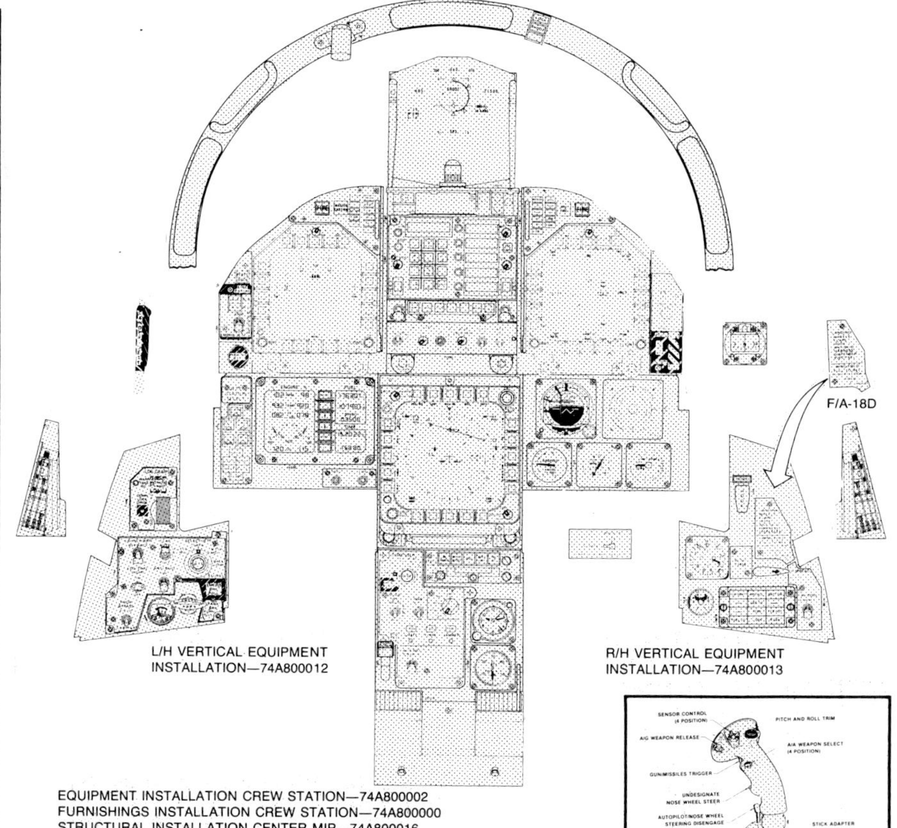

L/H VERTICAL EQUIPMENT INSTALLATION—74A800012

R/H VERTICAL EQUIPMENT INSTALLATION—74A800013

	ITEM	SOURCE	PART NUMBER	MAC DWG NUMBER	ELECT ITEM NO.
1	BRAKE PRESSURE INDICATOR	HTL	19999902-1	PS74-800066	8M-H052
2	L/H VERTICAL CONSOLE CONTROL PANEL	MCAIR		74A800724	52A-H077
3	EMERGENCY/PARK BRAKE CONTROL	MDEC-GRAND RAPIDS	7604D0001-02	74J808076	
4	LANDING GEAR CONTROL	MDAC-MONROVIA	717-102001-61	DS74-800073	12A-H008
5	INDICATOR PANEL	JAY-EL	51271-5	74J808065	52A-H084
6	INTEGRATED FUEL/ENGINE INDICATOR	GULL AIRBOURNE	372-047-002	PS74-870170	85A-H043
7	HDG & CRS CONTROL PANEL	MCAIR		74A800776	52AH098
8	MASTER ARM PANEL	MCAIR		74A800691	52A-H075
9	MULTIPURPOSE DISPLAY	KAISER AERO & ELECT	39000-69	PS74-870074	80A-H001
10	LEFT THREAT DISPLAY	JAY-EL	51703-110	74J808067	52A-H073
11	HEAD UP DISPLAY	KAISER AERO & ELECT	37000-69	PS74-870073	79A-J001
12	HUD VIDEO CAMERA	FAIRCHILD WESTON	1314A1	PS74-870137	
13	UP-FRONT CONTROL PANEL	MDEC—ST. CHARLES	A05A0227-4	DS74-870091	79A-J006
14	RIGHT THREAT DISPLAY	JAY-EL	51703-108	74J808067	52A-J074
15	MULTIPURPOSE DISPLAY	KAISER AERO & ELECT	39000-69	PS74-870074	80A-J002
16	MAP GAIN CONTROL	MCAIR		74A800693	52A-J076
17	ATTITUDE REFERENCE INDICATOR (ARU-48/A)	JET ELECTRONICS	501-1278-01	74J808057	33M-J015
18	AZIMUTH INDICATOR	APPLIED TECHNOLOGY	31-052173-02	(GFAE)	
19	STANDBY AIRSPEED INDICATOR (AVU-30/A)	KEARFLEX ENG	2160-M	74J808055	33M-J007
20	STANDBY ALTIMETER (AAU-39/A)	SMITHS INDUSTRIES	WL1650AM2	PS74-800054	33M-J002
21	VERTICAL SPEED INDICATOR (AVU-29/A)	KEARFLEX ENG	843-67-2A-M	74J808056	33M-J008
22	HORIZONTAL SITUATION DISPLAY	BENDIX	3775033-12	PS74-870078	80A-J003
23	CONTROL INDICATOR	APPLIED TECHNOLOGY	31-052176-01	(GFAE)	
24	ECM CONTROL PANEL	MCAIR		74A800828	52A-H087
25	RUDDER PEDAL ADJUSTMENT HANDLE	MDEC-GRAND RAPIDS	7608A0001-01	74J808084	
26	BUNO LIGHTPLATE		74B800070-143	74J808070	8DSJ017
27	CLOCK (ABU-24/A)	WALTHAM PRECISION	400257	74J808059	8M-J020
28	ALTIMETER, PRESS COMP'MENT (AAU-38/A)	KEARFLEX ENG	2109-5M	74J808058	8M-J021
29	ARRESTING HOOK CONTROL	MDAC-MONROVIA	716-102000-51	DS74-800077	19A-J003
30	HEIGHT INDICATOR	BENDIX CORP	3809413-3	PS74-870105	67M-J002
31	LANDING CHECKLIST LIGHTPLATE		74B800070-191	74J808070	8DSJ092
*31A	LANDING CHECKLIST LIGHTPLATE		74B800070-189	74J808070	8DSJ092
32	WINGFOLD CONTROL	MDEC-GRAND RAPIDS	7605B0027-01	74J808078	17A-J008
33	HYDRAULIC PRESSURE INDICATOR (AGU-15/A)	BENDIX CORP	3809414-2	74J808060	10M-J005
34	CAUTION LIGHT PANEL	MDAC-MONROVIA	715-101000-101	DS74-800069	8A-J042
35	STATIC SOURCE SELECTOR	MCAIR		ST7M473-1	
36	HANDLE CANOPY BOW	MCAIR		74A350719	
37	MIRROR	MCAIR		ST9M510-1	
38	XENON DISCHARGE TUBE INDICATOR	SYMBOLIC DISPLAYS	702182	PS74-800093	8DSJ150
*38A	XENON DISCHARGE TUBE INDICATOR	SYMBOLIC DISPLAYS	702180	PS74-800093	8DSJ150
39	L/H LOUVERS METAL - ECS DISTRIBUTION	MDEC-GRAND RAPIDS	7611D0001-01/-03	74J808062	
39A	R/H LOUVERS METAL - ECS DISTRIBUTION	MDEC-GRAND RAPIDS	7611D0001-02/-04	74J808062	
40	CANOPY JETTISON LEVER	OEA	2818300-111-02	PS74-800206	
41	STANDBY MAGNETIC COMPASS (AQU-3/A)	AIRPATH	CB-2125-T5	(GFAE)	33M-J009
42	KNEEBOARD CHART LIGHT	MCAIR		5M1693	8DSH143
43	AV COOL LIGHTPLATE		74B800070-439	74J808070	8DSJ025
44	CENTER LOUVER, ECS DISTRIBUTION	MDEC-GRAND RAPIDS	7617B0001-01	74J808062	

*FOR F/A-18D

EQUIPMENT INSTALLATION CREW STATION—74A800002
FURNISHINGS INSTALLATION CREW STATION—74A800000
STRUCTURAL INSTALLATION CENTER MIP—74A800016
EQUIPMENT INSTALLATION MIP—74A800028
EQUIPMENT INSTALLATION CANOPY ARCH—74A800025
AVIONICS COOLING INSTALLATION—74A830020
SECONDARY LIGHTING INSTALLATION—74A800675
SYSTEMS INSTALLATION—74A600111

L/H CONSOLE EQUIPMENT INSTALLATION—74800003
L/H CONSOLE STRUCTURAL INSTALLATION—74A800010
POWER QUADRANT INSTALLATION—74A800008

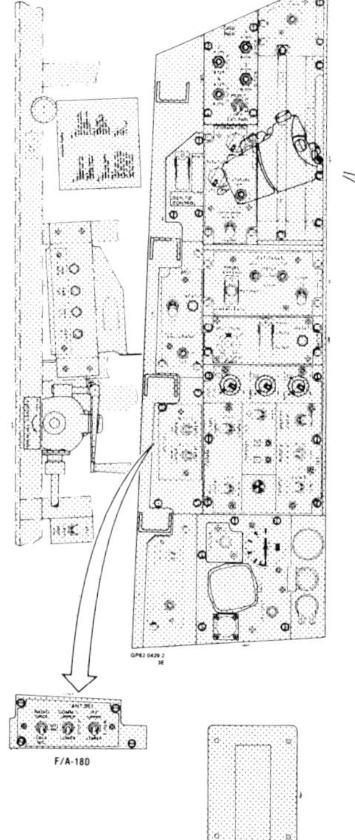

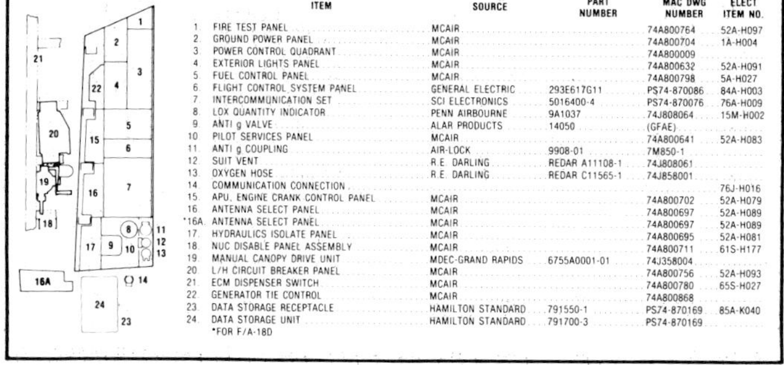

	ITEM	SOURCE	PART NUMBER	MAC DWG NUMBER	ELECT ITEM NO.
1	FIRE TEST PANEL	MCAIR		74A800764	52A-H097
2	GROUND POWER PANEL	MCAIR		74A800704	1A-H004
3	POWER CONTROL QUADRANT	MCAIR		74A800009	
4	EXTERIOR LIGHTS PANEL	MCAIR		74A800632	52A-H091
5	FUEL CONTROL PANEL	MCAIR		74A800798	5A-H027
6	FLIGHT CONTROL SYSTEM PANEL	GENERAL ELECTRIC	293E617G11	PS74-870086	84A-H003
7	INTERCOMMUNICATION SET	SCI ELECTRONICS	5016400-4	PS74-870076	76A-H009
8	LOX QUANTITY INDICATOR	PENN AIRBOURNE	9A1037	74J808064	15M-H002
9	ANTI g VALVE	ALAR PRODUCTS	14050	(GFAE)	
10	PILOT SERVICES PANEL	MCAIR		74A800641	52A-H083
11	ANTI g COUPLING	AIR-LOCK	9908-01	7M850-1	
12	SUIT VENT	R.E. DARLING	REDAR A11108-1	74J808061	
13	OXYGEN HOSE	R.E. DARLING	REDAR C11565-1	74J858001	
14	COMMUNICATION CONNECTION				76J-H016
15	APU, ENGINE CRANK CONTROL PANEL	MCAIR		74A800702	52A-H079
16	ANTENNA SELECT PANEL	MCAIR		74A800697	52A-H089
16A	ANTENNA SELECT PANEL	MCAIR		74A800697	52A-H089
17	HYDRAULICS ISOLATE PANEL	MCAIR		74A800695	52A-H081
18	NUC DISABLE PANEL ASSEMBLY	MCAIR		74A800711	61S-H177
19	MANUAL CANOPY DRIVE UNIT	MDEC-GRAND RAPIDS	6755A0001-01	74J358004	
20	L/H CIRCUIT BREAKER PANEL	MCAIR		74A800756	52A-H093
21	ECM DISPENSER SWITCH	MCAIR		74A800780	65S-H027
22	GENERATOR TIE CONTROL	MCAIR		74A800868	
23	DATA STORAGE RECEPTACLE	HAMILTON STANDARD	791550-1	PS74-870169	85A-K040
24	DATA STORAGE UNIT	HAMILTON STANDARD	791700-3	PS74-870169	

*FOR F/A-18D

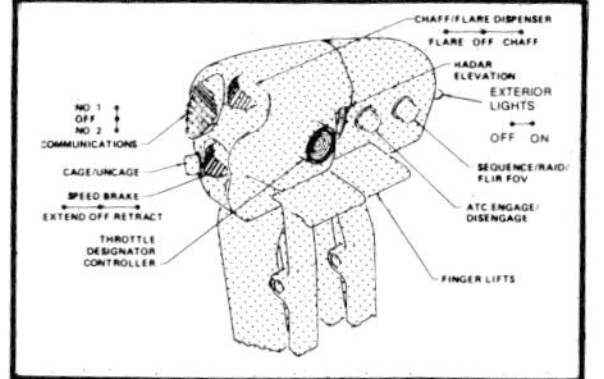

F/A-18C — F/A-18D
CREW STATION ORIENTATION

FORWARD COCKPIT
LEFT AND RIGHT VERTICAL CONSOLES

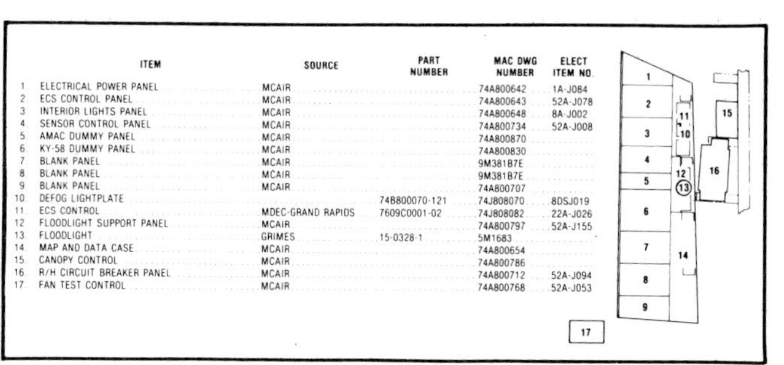

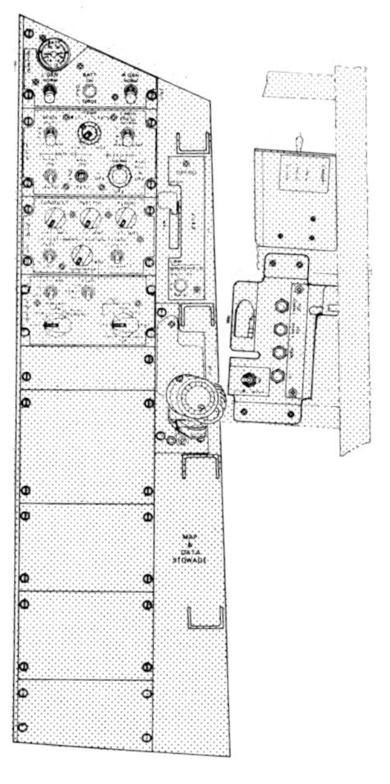

	ITEM	SOURCE	PART NUMBER	MAC DWG NUMBER	ELECT ITEM NO.
1	ELECTRICAL POWER PANEL	MCAIR		74A800642	1A-J084
2	ECS CONTROL PANEL	MCAIR		74A800643	52A-J078
3	INTERIOR LIGHTS PANEL	MCAIR		74A800648	8A-J002
4	SENSOR CONTROL PANEL	MCAIR		74A800734	52A-J008
5	AMAC DUMMY PANEL	MCAIR		74A800870	
6	KY-58 DUMMY PANEL	MCAIR		74A800830	
7	BLANK PANEL	MCAIR		9M381B7E	
8	BLANK PANEL	MCAIR		9M381B7E	
9	BLANK PANEL	MCAIR		74A800707	
10	DEFOG LIGHTPLATE		74B800070-121	74J808070	8DSJ019
11	ECS CONTROL	MDEC-GRAND RAPIDS	7609C0001-02	74J808082	22A-J026
12	FLOODLIGHT SUPPORT PANEL	MCAIR		74A800797	52A-J155
13	FLOODLIGHT	GRIMES	15-0328-1	5M1683	
14	MAP AND DATA CASE	MCAIR		74A800654	
15	CANOPY CONTROL	MCAIR		74A800786	
16	R/H CIRCUIT BREAKER PANEL	MCAIR		74A800712	52A-J094
17	FAN TEST CONTROL	MCAIR		74A800768	52A-J053

R/H CONSOLE EQUIPMENT INSTALLATION—74A800004
R/H CONSOLE STRUCTURAL INSTALLATION—74A800011

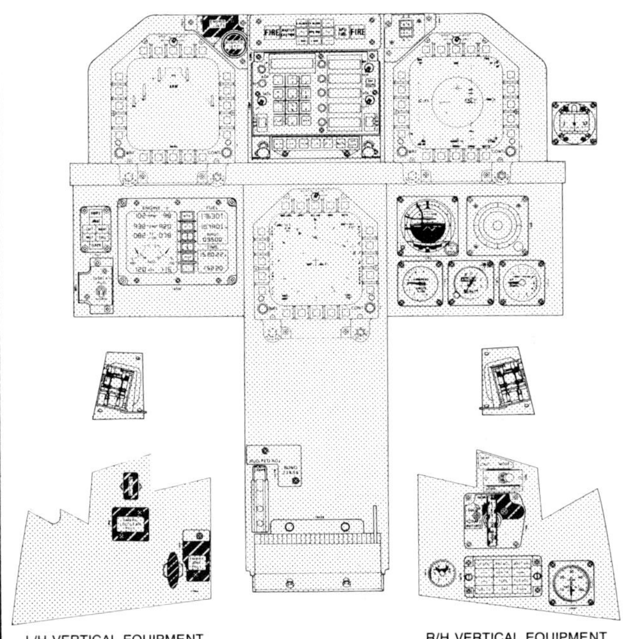

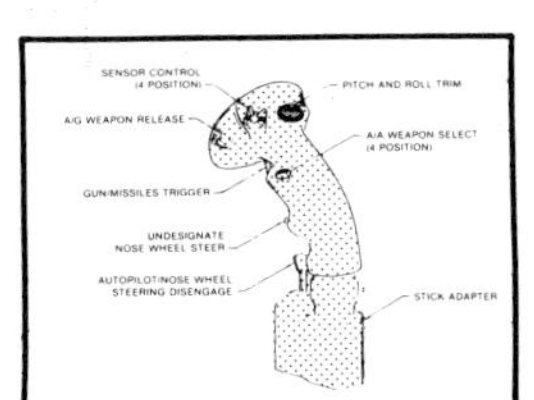

F/A-18D
CREW STATION ORIENTATION

AFT COCKPIT
MAIN INSTRUMENT PANEL
LEFT AND RIGHT VERTICAL CONSOLES

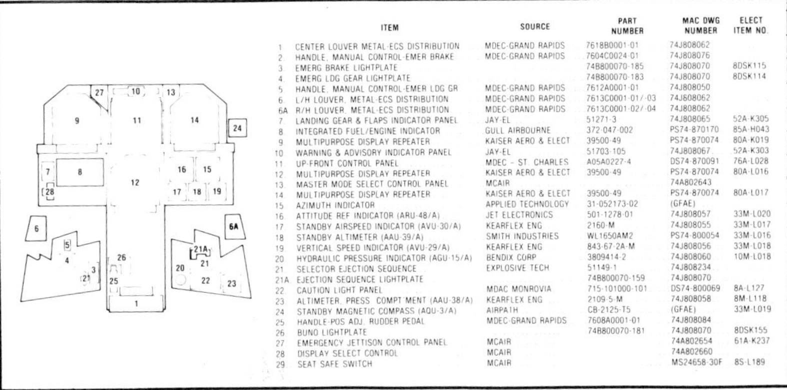

	ITEM	SOURCE	PART NUMBER	MAC DWG NUMBER	ELECT ITEM NO.
1	CENTER LOUVER METAL-ECS DISTRIBUTION	MDEC-GRAND RAPIDS	7618B0001-01	74J808062	
2	HANDLE, MANUAL CONTROL-EMER BRAKE	MDEC-GRAND RAPIDS	7604C0024-01	74J808076	
3	EMERG BRAKE LIGHTPLATE		74B800070-185	74J808070	8DSK115
4	EMERG LDG GEAR LIGHTPLATE		74B800070-183	74J808070	8DSK114
5	HANDLE, MANUAL CONTROL-EMER LDG GR	MDEC-GRAND RAPIDS	7612A0001-01	74J808050	
6	L/H LOUVER, METAL-ECS DISTRIBUTION	MDEC-GRAND RAPIDS	7613C0001-01/-03	74J808062	
6A	R/H LOUVER, METAL-ECS DISTRIBUTION	MDEC-GRAND RAPIDS	7613C0001-02/-04	74J808062	
7	LANDING GEAR & FLAPS INDICATOR PANEL	JAY-EL	51271-3	74J808065	52A-K305
8	INTEGRATED FUEL/ENGINE INDICATOR	GULL AIRBOURNE	372-047-002	PS74-870170	85A-H043
9	MULTIPURPOSE DISPLAY REPEATER	KAISER AERO & ELECT	39500-49	PS74-870074	80A-K019
10	WARNING & ADVISORY INDICATOR PANEL	JAY-EL	51703-105	74J808067	52A-K303
11	UP-FRONT CONTROL PANEL	MDEC - ST. CHARLES	A05A0227-4	DS74-870091	76A-L028
12	MULTIPURPOSE DISPLAY REPEATER	KAISER AERO & ELECT	39500-49	PS74-870074	80A-L016
13	MASTER MODE SELECT CONTROL PANEL	MCAIR		74A802643	
14	MULTIPURPOSE DISPLAY REPEATER	KAISER AERO & ELECT	39500-49	PS74-870074	80A-L017
15	AZIMUTH INDICATOR	APPLIED TECHNOLOGY	31-052173-02	(GFAE)	
16	ATTITUDE REF INDICATOR (ARU-48/A)	JET ELECTRONICS	501-1278-01	74J808057	33M-L020
17	STANDBY AIRSPEED INDICATOR (AVU-30/A)	KEARFLEX ENG	2160-M	74J808055	33M-L017
18	STANDBY ALTIMETER (AAU 39/A)	SMITH INDUSTRIES	WL1650AM2	PS74-800054	33M-L016
19	VERTICAL SPEED INDICATOR (AVU-29/A)	KEARFLEX ENG	843-67-2A-M	74J808056	33M-L018
20	HYDRAULIC PRESSURE INDICATOR (AGU-15/A)	BENDIX CORP	3809414-2	74J808060	10M-L018
21	SELECTOR EJECTION SEQUENCE	EXPLOSIVE TECH	51149-1	74J808234	
21A	EJECTION SEQUENCE LIGHTPLATE		74B800070-159	74J808070	
22	CAUTION LIGHT PANEL	MDAC MONROVIA	715-101000-101	DS74-800069	8A-L127
23	ALTIMETER, PRESS COMPT MENT (AAU-38/A)	KEARFLEX ENG	2109-5-M	74J808058	8M-L118
24	STANDBY MAGNETIC COMPASS (AQU-3/A)	AIRPATH	CB-2125-T5	(GFAE)	33M-L019
25	HANDLE-POS ADJ. RUDDER PEDAL	MDEC-GRAND RAPIDS	7608A0001-01	74J808084	
26	BUNO LIGHTPLATE		74B800070-181	74J808070	8DSK155
27	EMERGENCY JETTISON CONTROL PANEL	MCAIR		74A802654	61A-K237
28	DISPLAY SELECT CONTROL	MCAIR		74A802660	
29	SEAT SAFE SWITCH	MCAIR		MS24658-30F	8S-L189

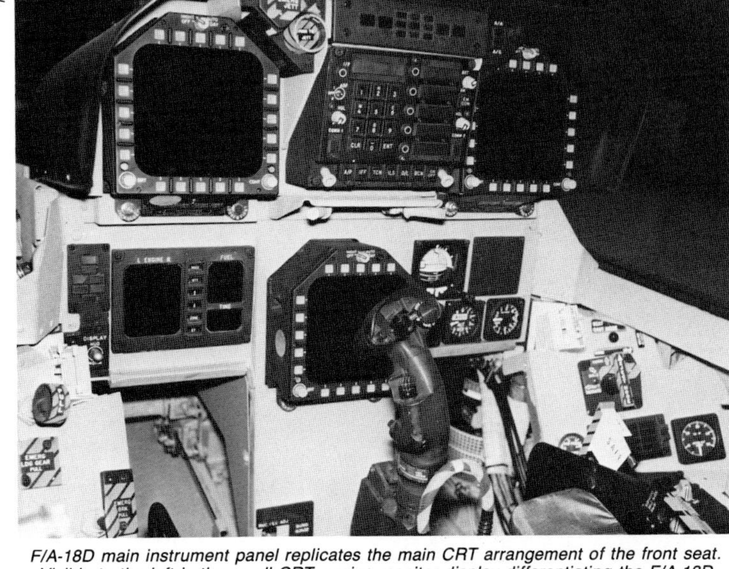

F/A-18D main instrument panel replicates the main CRT arrangement of the front seat. Visible to the left is the small CRT engine monitor display differentiating the F/A-18D from the F/A-18B. Control stick configuration is essentially unchanged.

F/A-18D
CREW STATION ORIENTATION

AFT COCKPIT
LEFT AND RIGHT CONSOLES

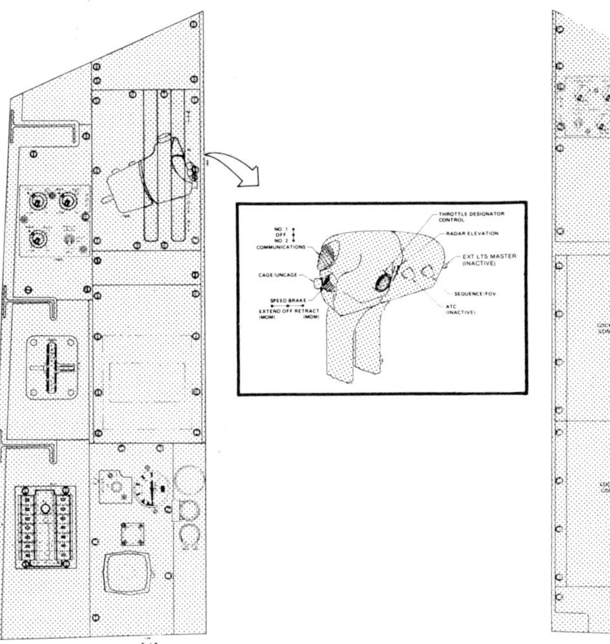

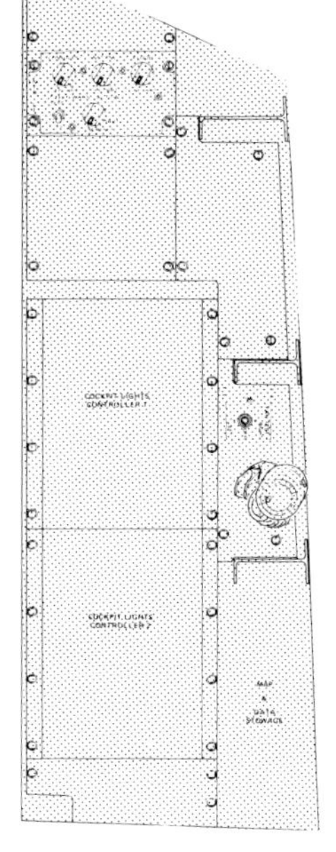

	ITEM	SOURCE	PART NUMBER	MAC DWG NUMBER	ELECT ITEM NO.
1	BLANK PANEL	MCAIR		74A802636	
2	POWER CONTROL QUADRANT	MCAIR		74A802053	
2A	THROTTLE CLOSURE PANEL	MCAIR		74A802638	
3	BLANK PANEL	MCAIR		9M381A1LE	
*4	AN/ARC-182 RADIO SET	COLLINS TELECOM	662-5663-002		77A-K001
4A	BLANK PANEL	MCAIR		9M381D7E	
5	PILOT SERVICES PANEL	MCAIR		74A802052	52A-H083
6	VOLUME CONTROL PANEL (ICS)	MCAIR		74A802632	76A-K032
7	CANOPY JETTISON CONTROL	OEA	2818200-103	PS74 800206	
8	ALE-39 CONTROL PANEL (ECM)	GOODYEAR AERO	MX-9254/ALE-39	(GFAE)	65A-K003
9	COMMUNICATION CONNECTION				76J-H016
10	ANTI g COUPLING	AIR—LOCK	9908-01	7M850-1	
11	SUIT VENT	R.E. DARLING	REDAR A11108-2	74J808061	
12	OXYGEN HOSE	R.E. DARLING	REDAR C11565-1	74J858001	

*UNDERNEATH BLANK PANEL

L/H CONSOLE EQUIPMENT INSTALLATION—74A802003
L/H CONSOLE STRUCTURAL INSTALLATION—74A802027
POWER QUANDRANT INSTALLATION—74A802015
SYSTEMS INSTALLATION—74A600120

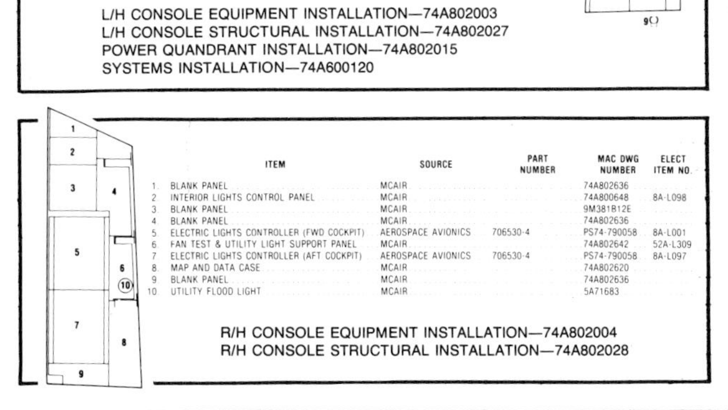

	ITEM	SOURCE	PART NUMBER	MAC DWG NUMBER	ELECT ITEM NO.
1	BLANK PANEL	MCAIR		74A802636	
2	INTERIOR LIGHTS CONTROL PANEL	MCAIR		74A800648	8A-L098
3	BLANK PANEL	MCAIR		9M381B12E	
4	BLANK PANEL	MCAIR		74A802636	
5	ELECTRIC LIGHTS CONTROLLER (FWD COCKPIT)	AEROSPACE AVIONICS	706530-4	PS74-790058	8A-L001
6	FAN TEST & UTILITY LIGHT SUPPORT PANEL	MCAIR		74A802642	52A-L309
7	ELECTRIC LIGHTS CONTROLLER (AFT COCKPIT)	AEROSPACE AVIONICS	706530-4	PS74-790058	8A-L097
8	MAP AND DATA CASE	MCAIR		74A802620	
9	BLANK PANEL	MCAIR		74A802636	
10	UTILITY FLOOD LIGHT	MCAIR		5A71683	

R/H CONSOLE EQUIPMENT INSTALLATION—74A802004
R/H CONSOLE STRUCTURAL INSTALLATION—74A802028

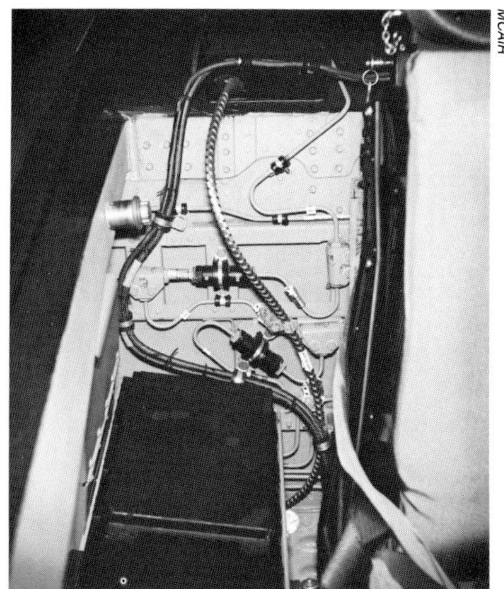

F/A-18D rear cockpit (starboard side) bulkhead supports miscellaneous systems lines, various electrical wiring bundles, and the arresting hook release cable.

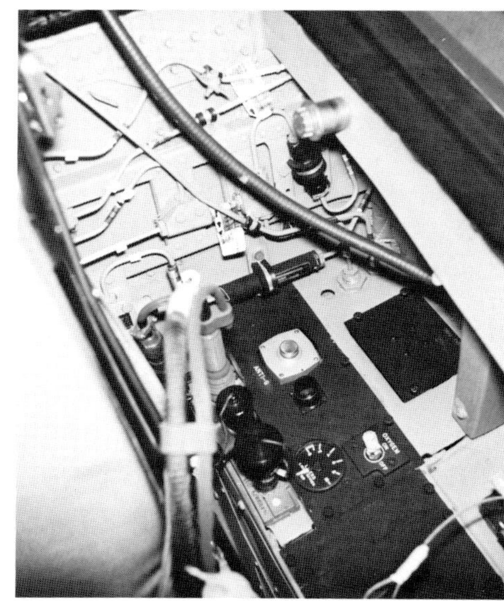

F/A-18D rear cockpit (port side) bulkhead and aft console supports the environmental control system and miscellaneous systems lines.

The F/A-18C (as well as the F/A-18A, F/A-18B, and F/A-18D) canopy is opened by an electro-pneumatically actuated mechanicl link tube assembly mounted behind the ejection seat. Visible in this view is the rubberized canopy seal designed to maintain pressurization. Massive locking hooks are noteworthy.

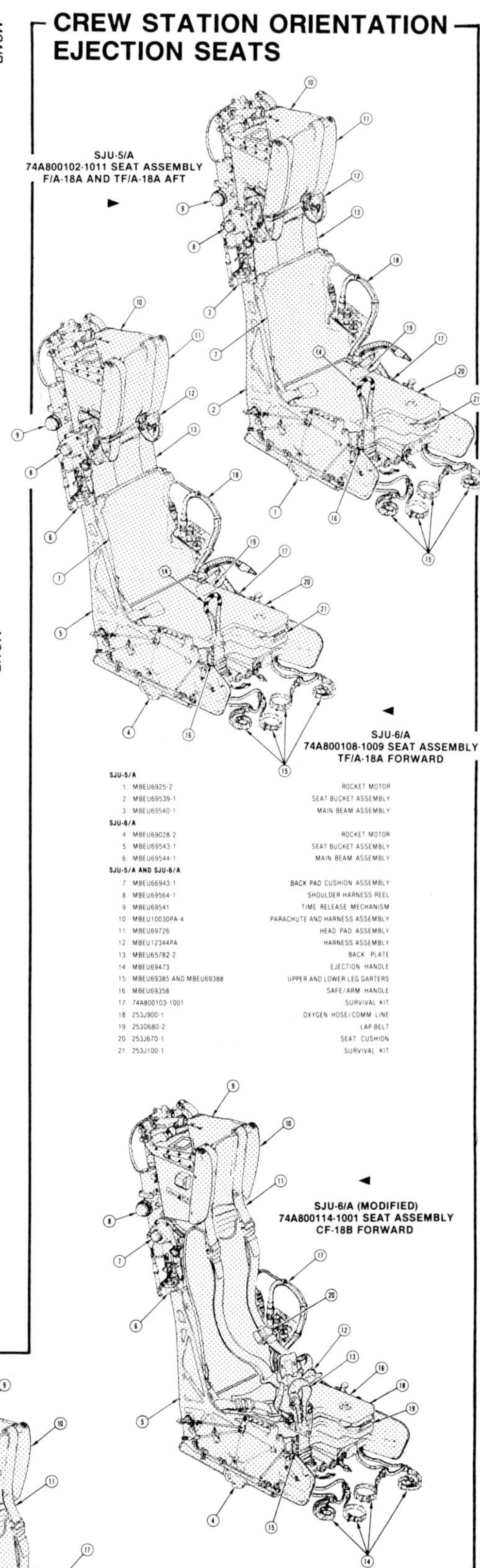

SJU-5/A

No.	Part	Description
1	MBEU6925-2	ROCKET MOTOR
2	MBEU69539-1	SEAT BUCKET ASSEMBLY
3	MBEU69540-1	MAIN BEAM ASSEMBLY

SJU-6/A

No.	Part	Description
4	MBEU69028-2	ROCKET MOTOR
5	MBEU69543-1	SEAT BUCKET ASSEMBLY
6	MBEU69544-1	MAIN BEAM ASSEMBLY

SJU-5/A AND SJU-6/A

No.	Part	Description
7	MBEU66943-1	BACK PAD CUSHION ASSEMBLY
8	MBEU69564-1	SHOULDER HARNESS REEL
9	MBEU69541	TIME RELEASE MECHANISM
10	MBEU10030PA-4	PARACHUTE AND HARNESS ASSEMBLY
11	MBEU69726	HEAD PAD ASSEMBLY
12	MBEU12344PA	HARNESS ASSEMBLY
13	MBEU65782-2	BACK PLATE
14	MBEU69473	EJECTION HANDLE
15	MBEU69385 AND MBEU69388	UPPER AND LOWER LEG GARTERS
16	MBEU69358	SAFE/ARM HANDLE
17	74A800103-1001	SURVIVAL KIT
18	253J900-1	OXYGEN HOSE/COMM LINE
19	253D680-2	LAP BELT
20	253J670-1	SEAT CUSHION
21	253J100-1	SURVIVAL KIT

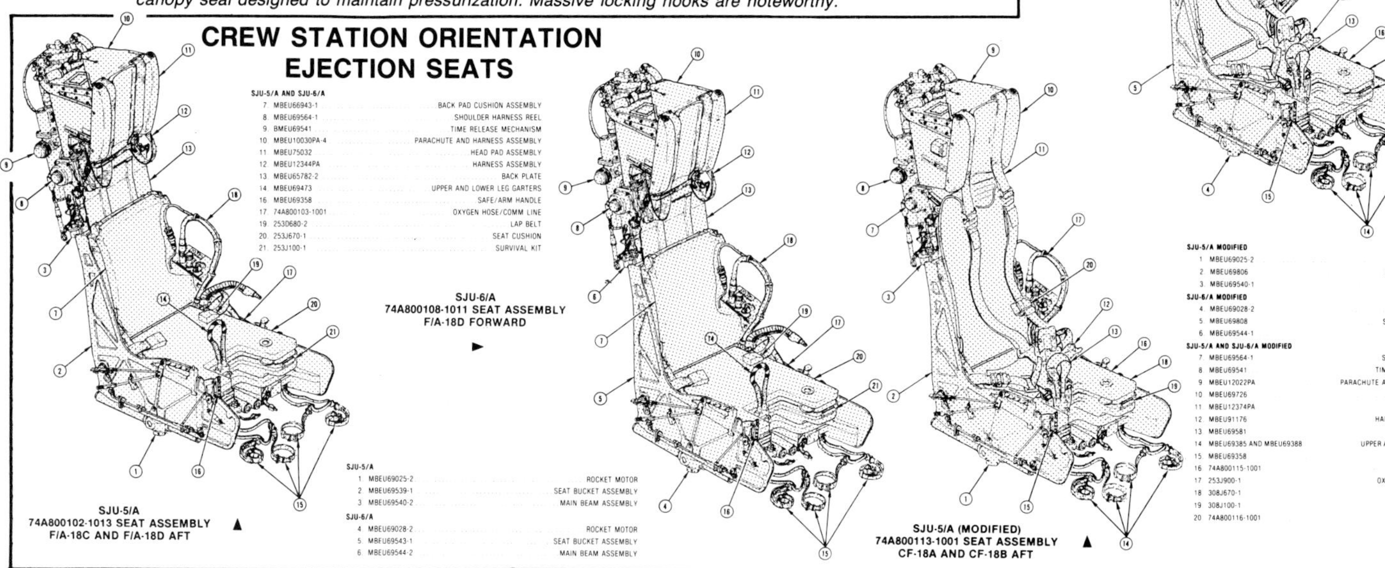

SJU-5/A AND SJU-6/A

No.	Part	Description
7	MBEU66943-1	BACK PAD CUSHION ASSEMBLY
8	MBEU69564-1	SHOULDER HARNESS REEL
9	BMEU69541	TIME RELEASE MECHANISM
10	MBEU10030PA-4	PARACHUTE AND HARNESS ASSEMBLY
11	MBEU75032	HEAD PAD ASSEMBLY
12	MBEU12344PA	HARNESS ASSEMBLY
13	MBEU65782-2	BACK PLATE
14	MBEU69473	UPPER AND LOWER LEG GARTERS
16	MBEU69358	SAFE/ARM HANDLE
17	74A800103-1001	OXYGEN HOSE/COMM LINE
19	253D680-2	LAP BELT
20	253J670-1	SEAT CUSHION
21	253J100-1	SURVIVAL KIT

SJU-5/A

No.	Part	Description
1	MBEU69025-2	ROCKET MOTOR
2	MBEU69539-1	SEAT BUCKET ASSEMBLY
3	MBEU69540-2	MAIN BEAM ASSEMBLY

SJU-6/A

No.	Part	Description
4	MBEU69028-2	ROCKET MOTOR
5	MBEU69543-1	SEAT BUCKET ASSEMBLY
6	MBEU69544-2	MAIN BEAM ASSEMBLY

SJU-5/A MODIFIED

No.	Part	Description
1	MBEU69025-2	ROCKET MOTOR
2	MBEU69806	SEAT BUCKET ASSEMBLY
3	MBEU69540-1	MAIN BEAM ASSEMBLY

SJU-6/A MODIFIED

No.	Part	Description
4	MBEU69028-2	ROCKET MOTOR
5	MBEU69806	SEAT BUCKET ASSEMBLY
6	MBEU69544-1	MAIN BEAM ASSEMBLY

SJU-5/A AND SJU-6/A MODIFIED

No.	Part	Description
7	MBEU69564-1	SHOULDER HARNESS REEL
8	MBEU69541	TIME RELEASE MECHANISM
9	MBEU12022PA	PARACHUTE AND HARNESS ASSEMBLY
10	MBEU69726	HEAD PAD ASSEMBLY
11	MBEU12374PA	HARNESS ASSEMBLY
12	MBEU91176	HARNESS RELEASE HANDLE
13	MBEU69581	EJECTION HANDLE
14	MBEU69385 AND MBEU69388	UPPER AND LOWER LEG GARTERS
15	MBEU69358	SAFE/ARM HANDLE
16	74A800115-1001	SURVIVAL KIT
17	253J900-1	OXYGEN HOSE/COMM LINE
18	308J670-1	SEAT CUSHION
19	308J100-1	SURVIVAL KIT
20	74A800116-1001	DELUTER DEMAND

Headrest of "Hornet" SJU-5/6 series ejection seats contains parachute and associated riser assembly and mounts scissor mechanism and manifold check valve.

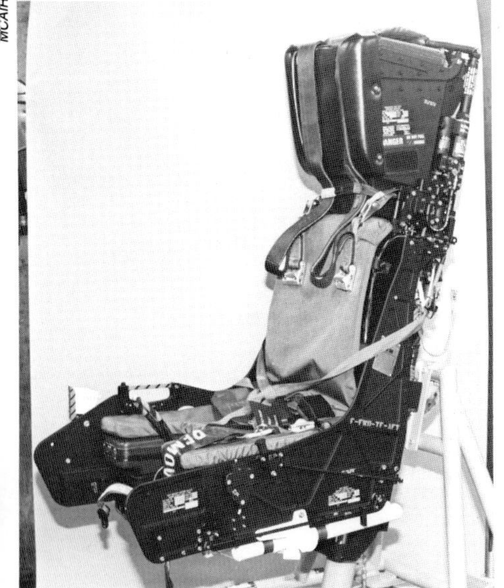

The SJU-5/6 series ejection seat is functional throughout almost all of the F/A-18s flight envelope and provides zero-zero emergency egress capability as well.

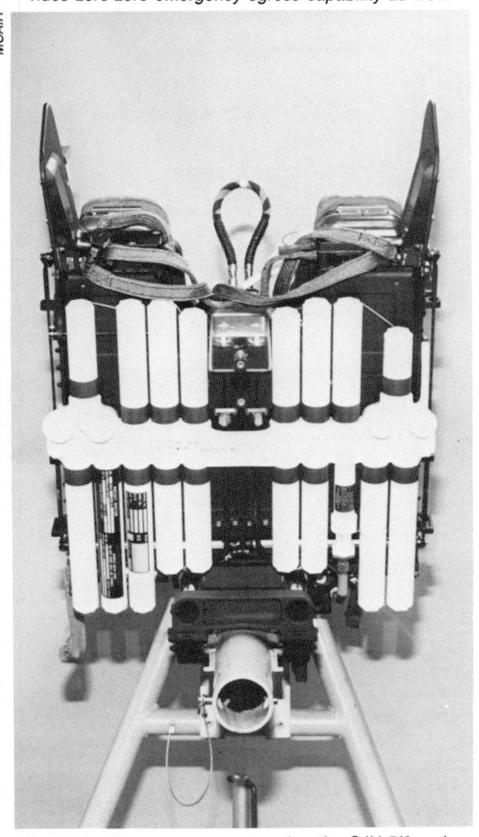

Propulsion for emergency egressing the SJU-5/6 series seat from the "Hornet" is provided by the tube-like rocket motor assembly mounted on the seat bottom.

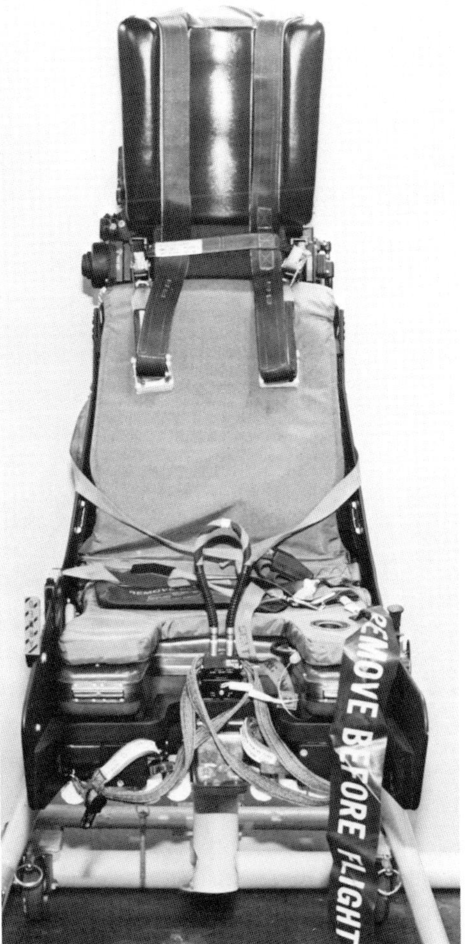

Front view of SJU-5/6 series ejection seat. Visible is seat padding, survival kit, ejection sequence initiation handle, headrest, and miscellaneous straps.

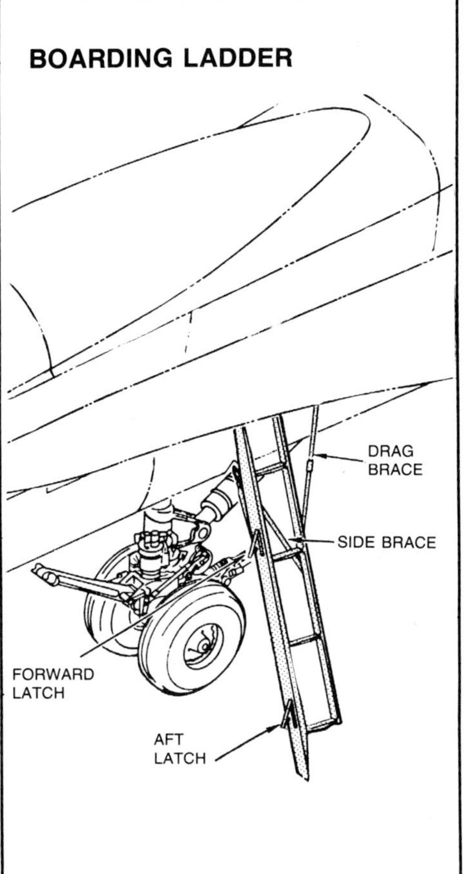

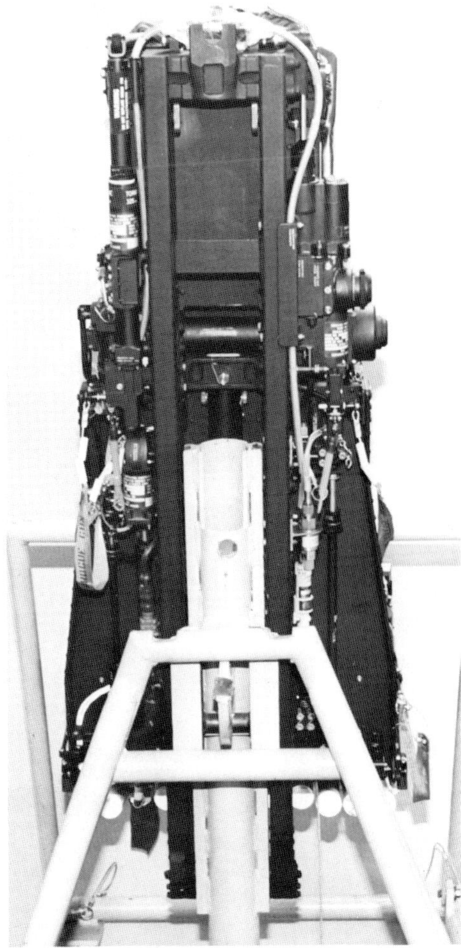

Rear view of SJU-5/6 series ejection seat. Visible is main beam assembly, pyrotechnic quick disconnect unit (on right beam), and the inertia reel.

The "Hornet" comes with a built-in ingress/egress ladder that mechanically retracts neatly into the under surface of the port leading edge extension.

With the ingress/egress ladder extended, a system of support tubes attach to the fuselage slide to provide necessary rigidity.

The "Hornet's" canopy and windscreen design are optimized to provide 360° of vision for the pilot. The windscreen creates virtually no distortion.

The "Hornet's" canopy is hinged at its aft end and opens vertically. Visible in this view of a CF-18A, is the searchlight seen only on CAF "Hornets".

The canopy for the F/A-18B/D is a two piece assembly with a single support hoop located between each piece of laminated acrylic. Note aft seat clearance.

F/A-18A nose mounts strip lights, AN/ALR-67(V) RHAW antenna fairing, ventral gun bay gas vents, the nose radome, and gun bay access panels.

CAF CF-18A nose varies little from its Navy counterpart with the exception of the night identification spotlight found on the aircraft port side.

Nose radome houses planar antenna for AN/APG-65 radar unit. Cutouts on top of aft end of radome are ports for rotary cannon.

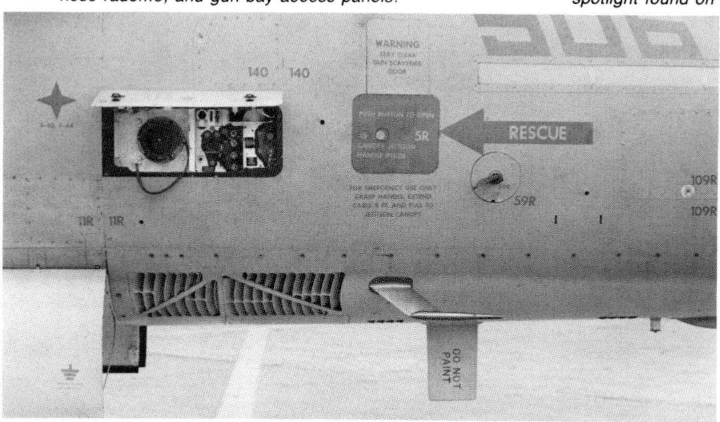

Right side of F/A-18A nose supports single-point refueling receptacle, a static pitot sensor, the emergency canopy release handle, and an angle of attack indicator system vane. Venting provides exit point for accumulated gun bay gases.

Left side of F/A-18A nose supports a static pitot sensor, a formation strip light, one AN/ALR-67(V) antenna fairing, gun and avionics bay venting, and an angle of attack indicator system vane. Additional EW system antenna are visible underneath nose.

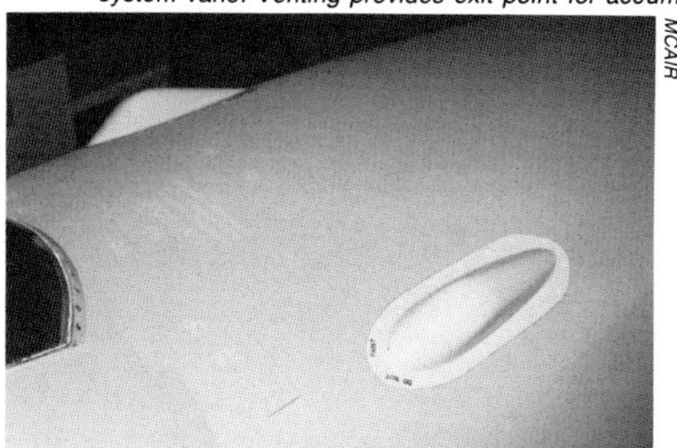

The F/A-18C/D series aircraft are equipped with eleven new or updated antennas. These are fairings for the new AN/ALQ-165 installation behind the canopy.

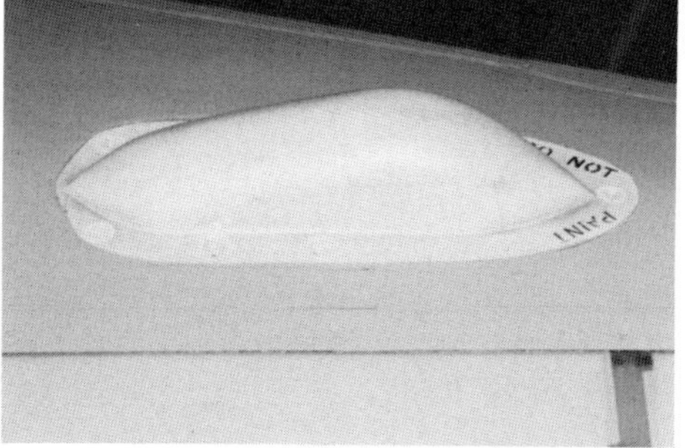

Part of the F/A-18C/D EW and weapons system related update includes two AN/ALQ-165 antenna and their associated fairings protruding from the nose top side.

The AN/ALR-67 antennas are provided fairings on either side of the nose just below and ahead of the most forward point of each leading edge extension.

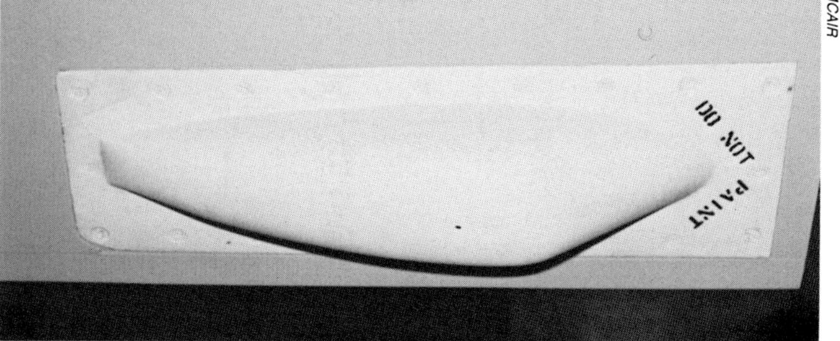

Exterior of the F/A-18C/D gun bay access door showing, from left to right, the AN/ALR-67 antenna (five prongs), the radar beacon antenna, and the AN/ALQ-165 high band transmission antenna fairing.

The AN/ALQ-165 low-band transmission antenna fairing is located on the F/A-18C's forward nose landing gear door. "Do not paint" nomenclature is often found on dielectric fairings of this nature due to the metallic content of most paints.

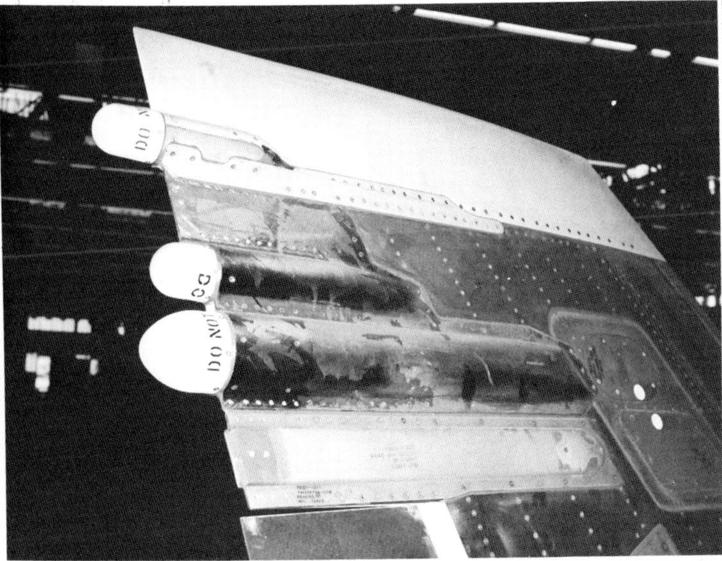

F/A-18C/D port vertical tail inside surface features (from top) the AN/ALQ-165 high-band transmission antenna (port fin only), the AN/ALR-67 antenna (both fins), a receiver antenna (port fin only), and a fuel dump nozzle.

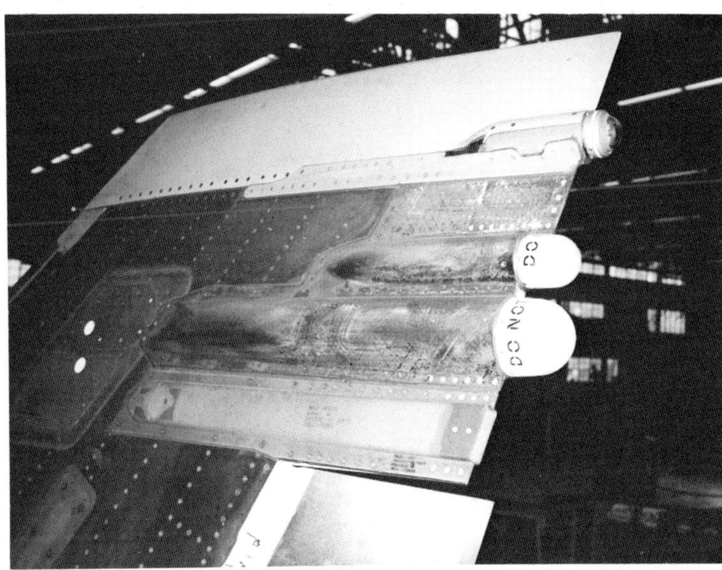

F/A-18C/D starboard vertical tail inside surface features (from top) the tail position light (starboard fin only), an AN/ALR-67 antenna fairing (both fins), an AN/ALQ-165 low-band transmission antenna fairing (starboard only), and a fuel dump nozzle.

Short-lived dog-tooth-style outboard panel wing leading edge extension is readily discernible in this view taken from the cockpit of what apparently is BuNo. 160775.

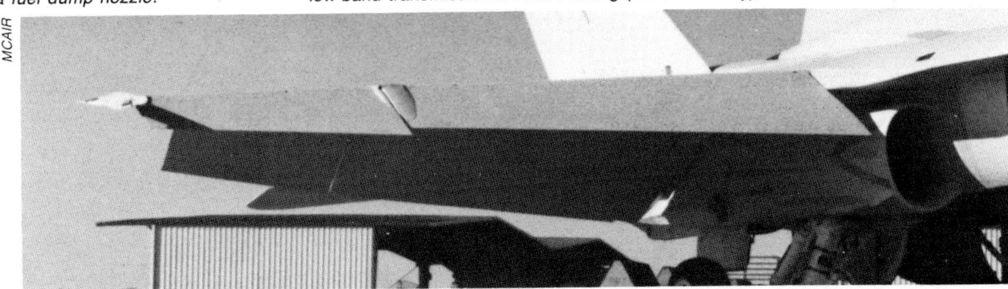

The "Hornet's" leading edge flap displacement has increased from an initial angle (shown) of 30° to the current angle of 35°. The increased angle is the result of a need to lower carrier approach speeds and cockpit forward viewing angles. BuNo. 160785, shown, is the 9th single-seat FSD aircraft.

A hydraulically actuated wing-fold mechanism moves the "Hornet's" outer wing panels approximately 10° past the vertical during the fold sequence.

The F/A-18C/D trailing edge flap hinge assemblies are external to the wing and faired at the hinge point by small teardrop-shaped end plates.

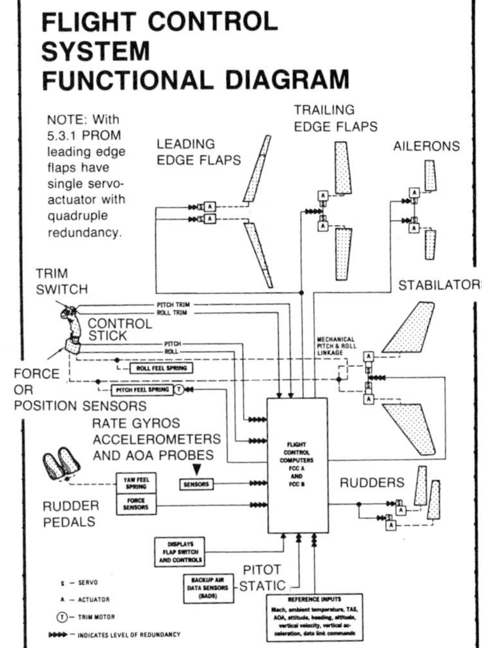

Poor roll rate and approach speed performance dictated a redesign of several F/A-18 components, including the trailing edge flaperons. The original, small flaperons can be seen on the left. The newer, increased area versions are on the right.

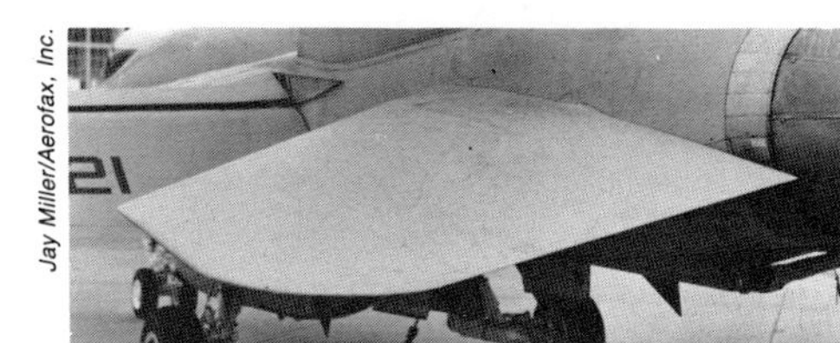

The "Hornet's" trailing edge flap and flaperon design has evolved into a two-section, slotted unit that functions in conjunction with the leading edge via the control system computers. The flaperons combine the roll control features of conventional ailerons with the lift/drag features of flaps.

The "Hornet's" primarily composite construction slab stabilators provide both pitch and roll control throughout the aircraft's flight envelope.

The vertical tail surfaces are canted outward from the aircraft's vertical axis. Each mounts a single-piece triply-hinged hydro-mechanically actuated rudder.

The vertical tail surfaces have been plagued with a series of fatigue anomalies which have been rectified by beefed up structure and external bracing.

The emergency fuel dump system nozzles are located just above the top edge of each rudder. High placement keeps dumped fuel out of exhaust plume.

Engine exhaust nozzles, vertical fins, and slab stabilators are cleanly faired on the empennage section—an area notorious for generating high drag.

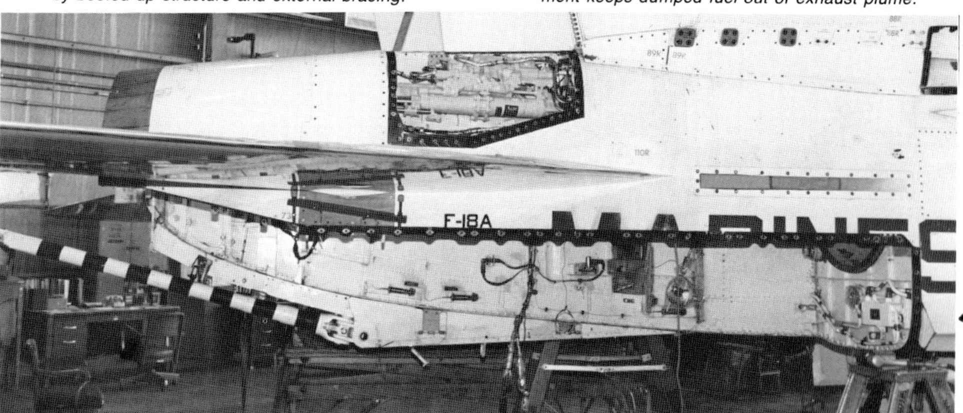

The slab stabilator hydraulic actuators are mounted inside the empennage fairing area located just below each rudder. The rams move the stabilators independently via a link arm assembly connected to the load-bearing carry-through spindle sleeve. Noteworthy is tail hook keel attachment point.

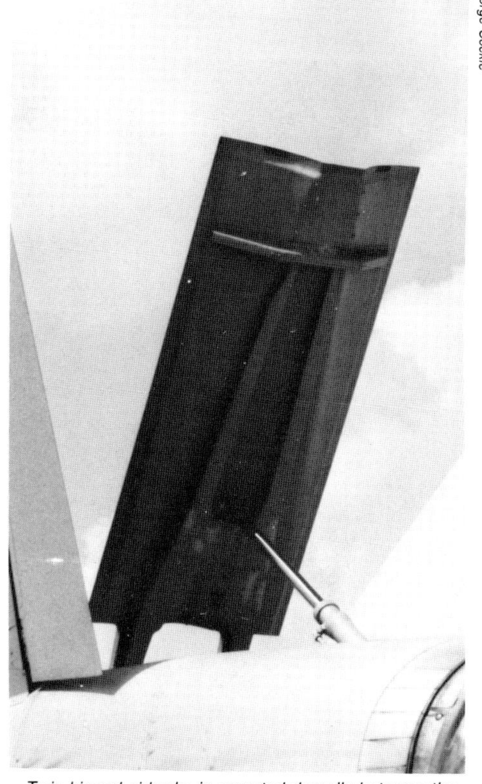

Twin-hinged airbrake is mounted dorsally between the vertical tail surfaces and is hydraulically actuated upon pilot command by a single hydraulic ram.

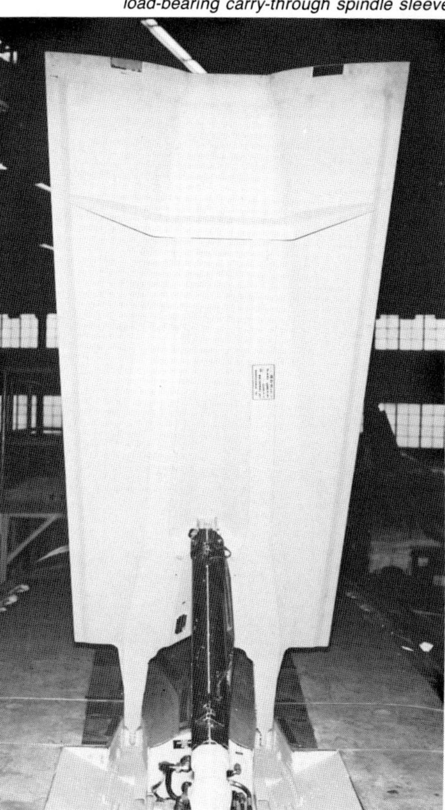

The airbrake is essentially a single-piece surface of graphite epoxy composite construction. Safety lock collar is visible around hydraulic actuator.

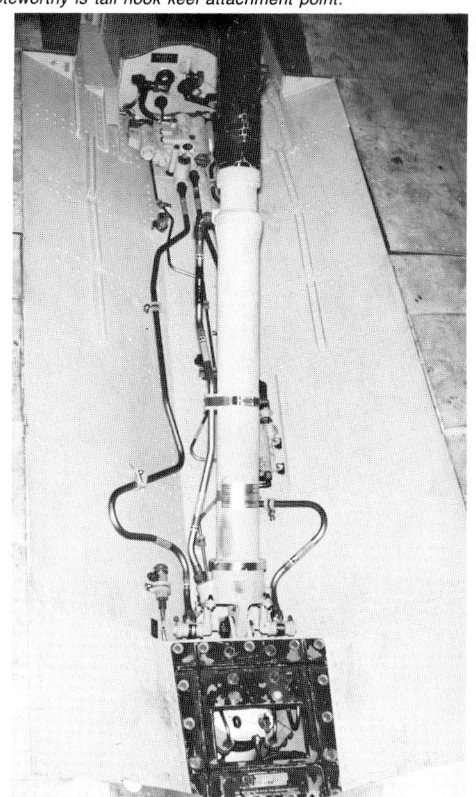

When retracted, airbrake fits flush against empennage section upper surface. Hydraulic actuator and additional panel (removed) fits inside available space.

Hook extension is by gravity and a nitrogen charged actuator. To ensure cable engagement and alleviate skipping tendencies, a retraction actuator cylinder damper provides the down force to hold the hook in proper position.

The hook is attached to the main aircraft keel assembly between the engine exhaust nozzles and thus is securely anchored to accommodate the excessive loads generated during carrier landings. Hooks are expendable and thus periodically replaced.

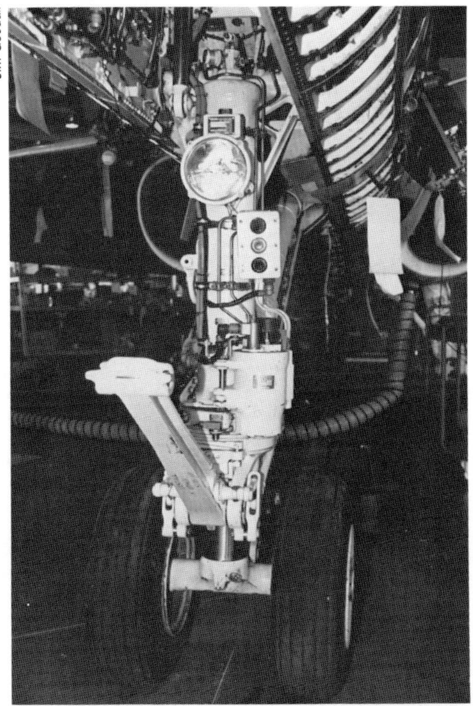

The nose landing gear retracts forward into its well. Hydraulic lines, actuators, and wire harnesses take up most of the available gear well space.

Early F/A-18A nose gear (FSD a/c) varies only slightly in configuration from that of production aircraft. Shuttle link bar is in its "up" position.

F/A-18A shuttle link bar actuator assembly is attached directly to the main strut assembly. Upon pilot command, it moves the bar up or down.

F/A-18A early nose gear (FSD a/c) seen from starboard side. Anti-torque scissor and link bar actuator assembly are to the rear of main strut.

Standard F/A-18A nose gear has revised anti-torque scissor hinge section. Retraction strut is long and robust to accommodate extraordinary loads generated during carrier takeoffs. Noteworthy is tight clearance between retraction strut and nose of centerline tank.

Port side of early F/A-18A nose gear assembly. Link bar is moved into position once the aircraft is properly positioned over the catapult shuttle.

The nose gear retraction strut is equipped with its own, integral fairing. As the gear rotates forward, a total of four well doors, including the retraction strut's integral fairing, are utilized. The forward doors (not visible here) are asymmetric in size and placement in concert with the well design.

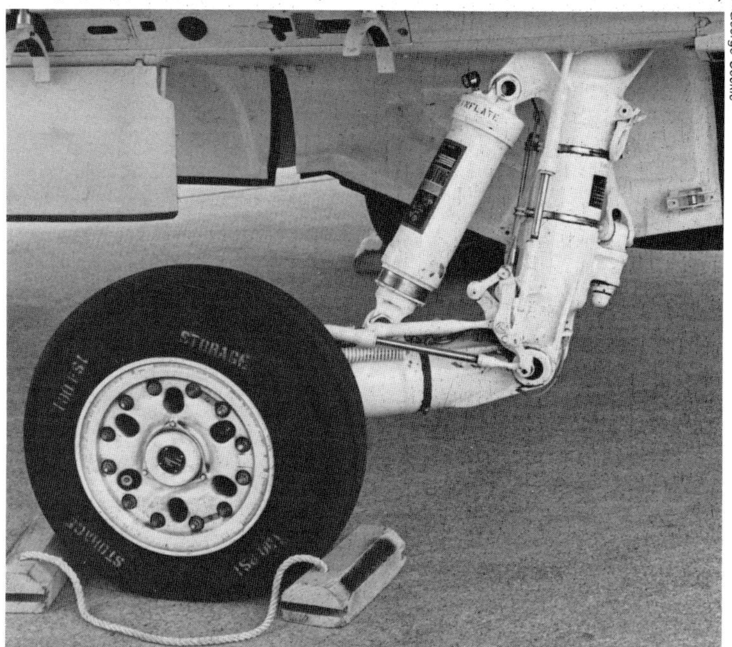

Starboard main gear illustrating trailing arm design. Early shock strut, with short stroke is shown. Higher than expected landing load dictated a shock strut redesign following initial carrier trials and operations.

Newer shock strut, seen on port F/A-18A main gear offers longer stroke and consequently greater vertical load tolerance. Hydraulic rams rotate the main gear some 90° during the retraction sequence into their respective wells.

The main landing gear wheels have hydraulically actuated disc brakes operated by pilot toe action on the rudder pedals. A computerized anti-skid system, automatically disabled at speeds below 30 kts., prevents wheel lock-up.

Maximum main gear loads, generated primarily during landings, are transmitted to the main airframe structure through main load-bearing bulkheads. Fatigue anomalies have developed in several of these bulkheads and fixes currently are being studied.

The main gear retract aft while rotating 90° to fit into main gear well. Retraction is hydraulically accommodated via rams in the gear well.

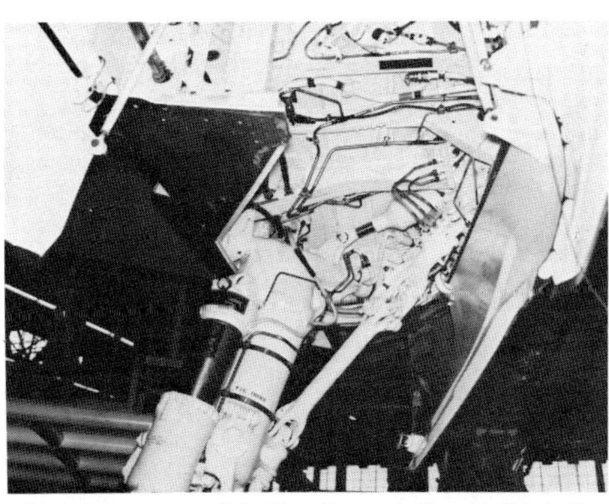

Port main gear well looking forward. Gear well interior is painted white so that oil and hydraulic leaks can be more readily visually identified.

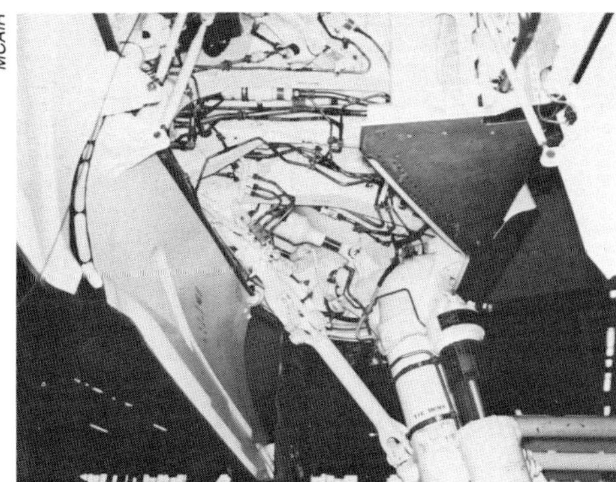

Starboard main gear well looking forward. Hydraulic actuating ram is visible near the center with bracing strut running diagonally from gear strut to well.

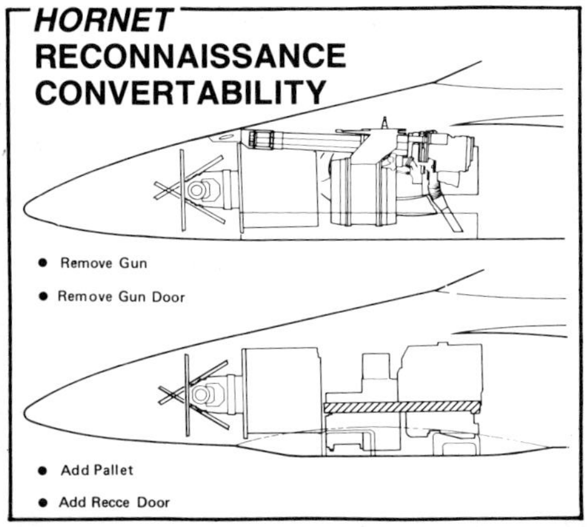

HORNET RECONNAISSANCE CONVERTABILITY

Port main gear well looking aft. Space serves as mounting point for various filters, electrical harnesses, and hydraulic plumbing.

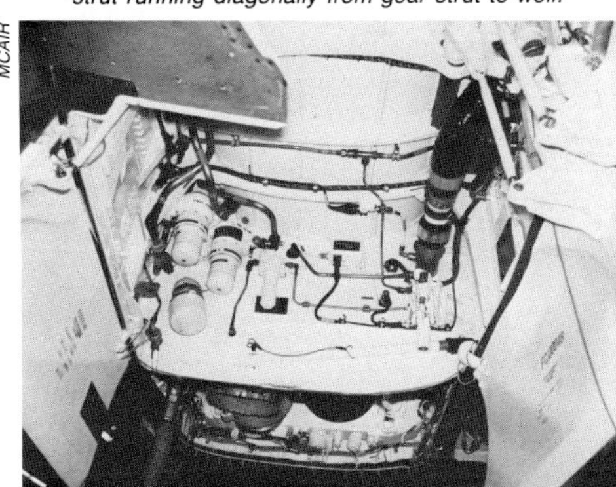

Starboard main gear well looking aft mirrors that of the port side well with filters, electrical harnesses, and hydraulic plumbing mounted on rear bulkhead.

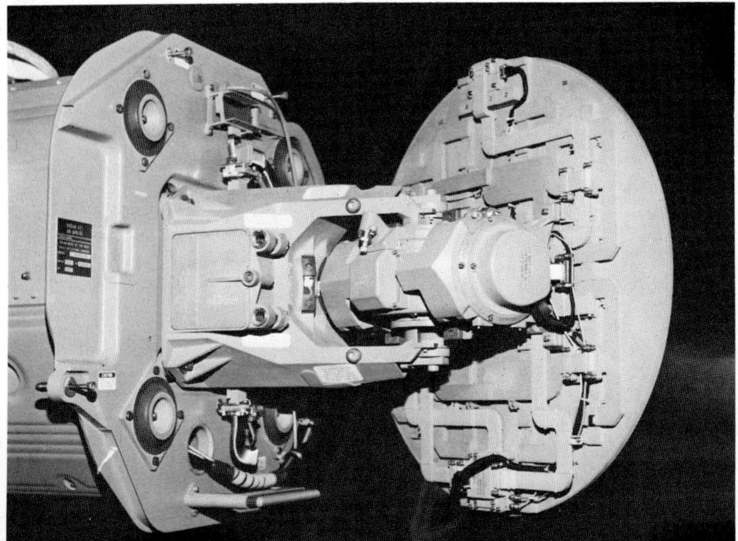

The relatively small and light Hughes AN/APG-65 is a conventional coherent pulse-Doppler radar operating in X-band (8 to 12.5 GHz). It has a programmable signal processor which performs up to 7.2 million operations per second.

The Hughes AN/APG-65 is rail mounted to permit easy access for maintenance. With the radome open, the entire unit slides out. The antenna is electrically actuated and articulated in both azimuth and elevation.

In the air-to-air mode, the Hughes AN/APG-65 provides velocity search, range-while-search, track-while-scan, raid assessment, gun director mode, and three missile attack modes (boresight, vertical acquisition, and head-up display).

The flat plate antenna is the largest that can be accommodated within the confines of the "Hornet's" nose radome dimensional constraints. The latter were dictated by aerodynamic requirements and mission objectives.

The leading edge extension, as well as improving aerodynamic characteristics, also helps accommodate intake flow requirements, particularly at high angles of attack. Intake splitter plate separates stable flow from unstable boundary layer air.

The "Hornet's" fixed ramp intake is purposefully mechanically uncomplicated and optimized to meet engine airflow needs throughout the aircraft's flight envelope. Boundary layer bleed slots are visible as vertical gray bars on splitter plate.

Latest mod, to be incorporated in the production line and retrofitted to all extant "Hornets", is dorsal fin on LEX to alleviate vertical tail aerodynamic loads.

The intake lip is curvilinear and sufficient in assimilating shock formation to adequately serve the needs of the "Hornet's" F404s up to about Mach 1.95.

Behind the splitter plate is the ram air intake for the air-conditioning and cabin pressurization system. Splitter plate is offset approximately 8 in.

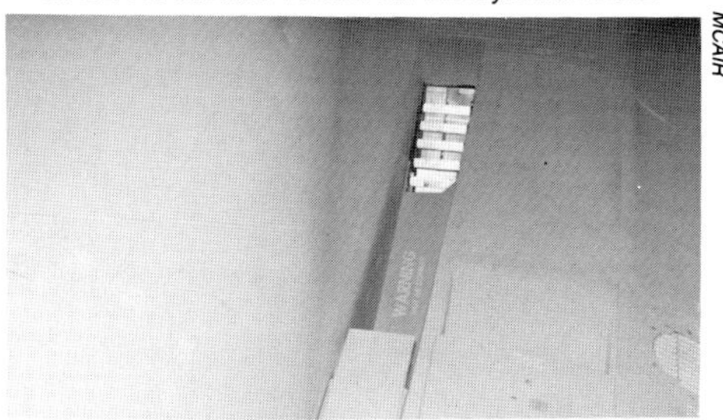

Boundary layer bleed system outlet doors are mounted dorsally in the spacer mounted between the intake splitter plate and the fuselage.

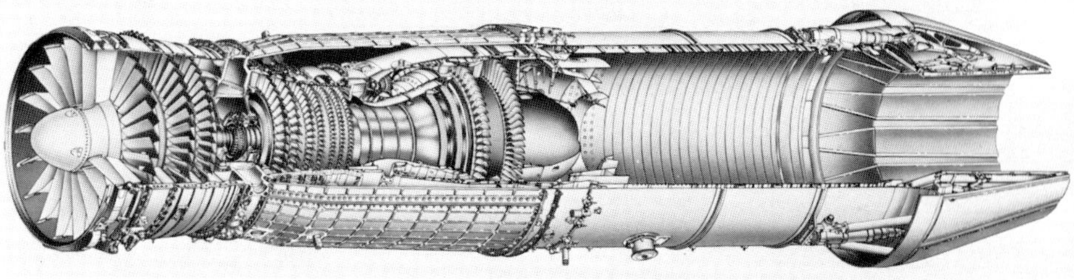

The General Electric F404-GE-400 is a low-bypass ratio turbofan optimized to provide a high thrust-to-weight ratio, good fuel consumption specifics, and high reliability. Unusually, accessories are mounted on the airframe rather than the engine, thus eliminating the need to develop separate left- and right-handed configurations.

The F404-GE-400's compressor is equipped with broad chord blades. Total cross sectional frontal area is extremely low for an engine in its power class.

The F404-GE-400 consists of seven basic modules which are optimized to facilitate ease of maintenance. When repair is required, the engine is suspended vertically rather than horizontally so that the faulty module can be removed without total engine disassembly. Simplicity of exhaust nozzle actuator assembly is noteworthy.

The F404-GE-400 can be removed from the "Hornet" in less than thirty minutes. A four-man crew is all that is required to pull an engine for repair or replacement. Large engine bay doors under the "Hornet's" aft fuselage open inward toward the aircraft centerline; the engine then is lowered out of the engine bay.

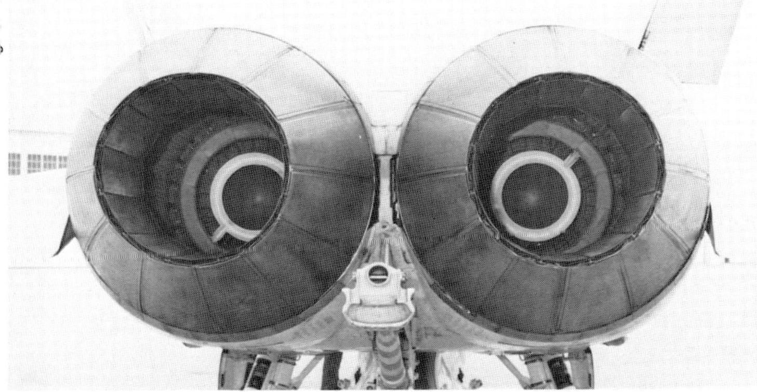

The F404-GE-400's exhaust nozzles are neatly faired into the aft end of the ''Hornet's'' fuselage, thus minimizing drag in an area that is historically high drag in nature. The exhaust nozzles dilate to accommodate varying exhaust gas flow requirements.

The inflight refueling probe with standardized connector is mounted on the starboard side of the nose, just ahead of the windscreen. It is hydraulically actuated and equipped with its own well fairing and a small well door.

All load bearing structures in the engine bay are located in what is essentially the bay's top half. This frees the lower half of obstructions and thus permits the rapid removal of an engine for repair or maintenance.

A hinge-like actuator moves the inflight refueling probe into and out of its well. This device, in turn, is moved by a small hydraulic ram.

TANK	USABLE FUEL		
	GALLONS	POUNDS JP-5	POUNDS JP-4
Number 1	418	2,840	2,720
Number 2 Left Engine Feed	263	1,790	1,710
Number 3 Right Engine Feed	206	1,400	1,340
Number 4	532	3,620	3,460
Total Fuselage	1,419	9,650	9,230
Left and Right Internal Wings	85 85 170	580 580 1,160	550 550 1,100
Total Internal	1,589	10,810	10,330
EXTERNAL TANK(S)			
Elliptical Wing or Centerline Tank	314	2,140	2,040
Cylindrical Wing or Centerline Tank	330	2,240	2,150

FUEL QUANTITY

NOTES

- The fuel quantities, in pounds, are rounded off to the nearest 10 pounds. Therefore, the actual gallons times 6.8 or 6.5 will not necessarily agree with the pounds column.

- Fuel weights are based on JP-5 or JP-4 at 6.8 or 6.5 pounds per gallon and a temperature of 65°F.

FUEL SYSTEM

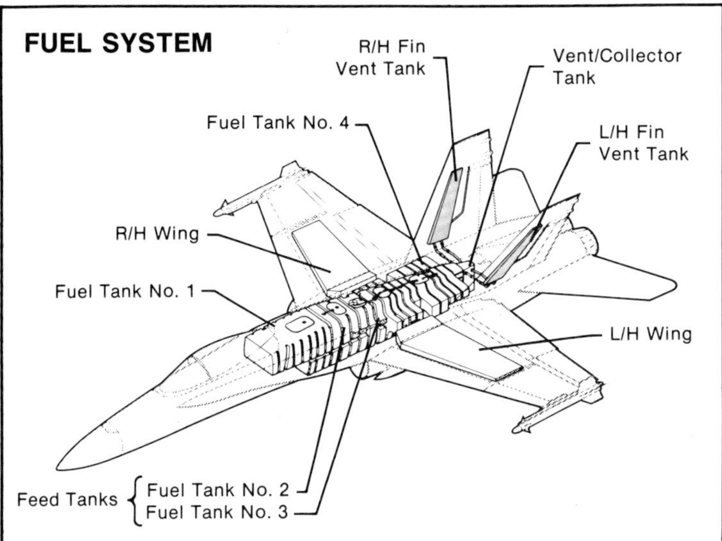

Refueling the ''Hornet'' in flight is simplified by the near-ideal placement of the probe and the pilot's ability to see it and the basket as the two come together for hook-up. Slight off-set permits reasonable visibility forward.

The standard drop tank for the "Hornet" can be used on either of its two inboard wing stations or the aircraft's centerline pylon—or all three. In any of the three positions the tank can be jettisoned upon pilot command. Tank capacity is 330 gals.

When the drop tank is attached to a wing pylon, its aft pivot attachment fitting is also utilized. This small device facilitates clean release of the tank by forcing it to rotate nose down, first.

Eight 500 lb. Mk.82 iron bombs are seen hanging from four horizontal ejector racks which, in turn, are suspended from four wing pylons of a VMFA-323 F/A-18A. This load represents less than one-fourth what the "Hornet" can carry at gross weight.

Part of the weaponry selection the "Hornet" is capable of carying. Noticeably absent are nuclear weapons. See p. 46 of this monograph for a more detailed listing of the "Hornet's" armament options and a description of those pictured here.

The original "Hornet"-optimized external fuel tank design was oval in shape. Premature fatigue anomalies, directly the result of the oval configuration, forced MCAIR to go to a more conventional, round tank design.

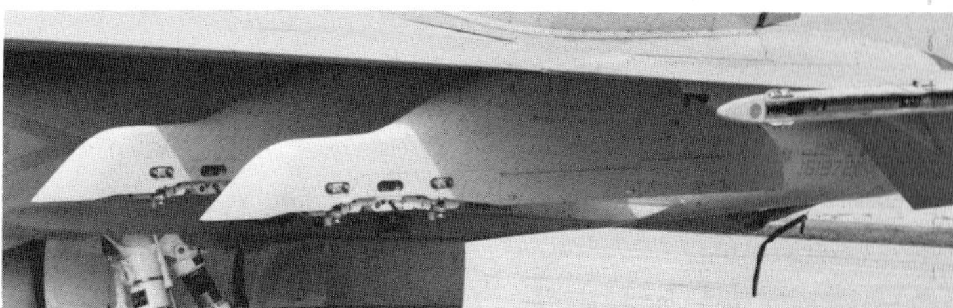

The "Hornet"-dedicated wing pylons are optimized to carry a wide variety of fuel and ordnance needs. Various rack configurations can be attached to each which, in turn, permit great variations in weaponry type.

The pylon's attachment mechanism is faired into the basic pylon shell and is suitable for a variety of triple ejector racks, multiple ejector racks, and horizontal ejector racks. Pylon weight capacity varies, but loaded drop tanks weigh 2,816 lbs.

Wingtip rails are designed to accommodate only the AIM-9L and AIM-9M "Sidewinder" infrared AAM. The rails are angled downward slightly out of consideration for the "Hornet's" cruising flight attitude.

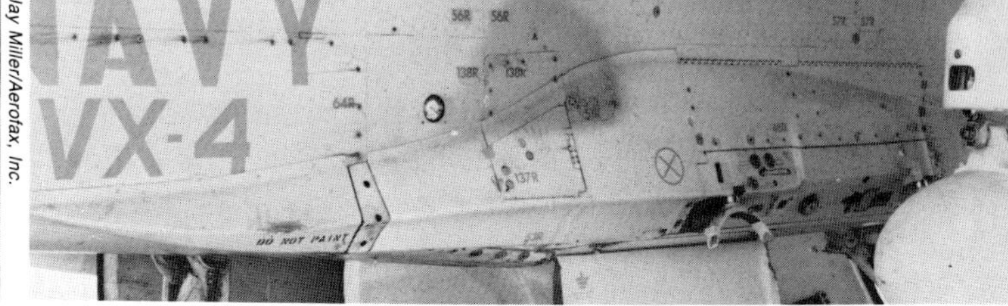

The AIM-7F radar-guided AAM attachment fairing and ejector arms are integral to each engine nacelle. Each fairing accommodates one missile or a single specialized sensor pod such as the LST/SCAM (starboard) or FLIR (port). The aft end of the AIM-7F attachment fairing accommodates the aft mid-band antenna and fairing for the AN/ALQ-126. A bump fairing underneath the intake accommodates the forward mid band AN/ALQ-126 antenna. Visible just ahead of main gear strut in left photo is flare dispenser.

Two versions of the "Sidewinder" infrared guided AAM normally are capable of being carried by the "Hornet". These include the AIM-9L (shown) and the AIM-9M. AIM-9s are carried on the wingtip rails and wing pylon (when rail equipped).

Six AIM-9Ls and two AIM-7Fs are being carried by a VMFA-531 F/A-18A. Four of the AIM-9s appear to be mounted on the aircraft's outboard wing pylons, indicating that room remains for the addition of two more pylons and their respective paylods.

AIM-7F is partially faired into the side of the engine nacelle when attached. The main landing gear, as they retract, go down and then underneath the missile's forward, lower fin before moving up into the main gear well.

The AGM-88A "HARM", seen on the starb'd outboard wing pylon, was utilized during "Operation El Dorado Canyon" (the punitive air strikes against Libya). The centerline store is a CBU-59/B "Rockeye" and the port outboard wing pylon store is an AIM-7F.

An F/A-18A of VMFA-323, armed with at least one AGM-88A "HARM" under its starb'd wing is seen departing the USS "Coral Sea" during "Operation El Dorado Canyon". A single AIM-7F is visible under the port wing and each wingtip mounts a single AIM-9L.

The McDonnell Douglas AGM-84 "Harpoon", seen mounted under the starboard wing of the prototype F/A-18C, BuNo. 163427. This is a relatively large and heavy weapon, thus limiting other external stores to those of light weight.

The AGM-109 "Harpoon" anti-ship missile is one of the lesser known members of the "Hornet's" weapon repertoire. Though heavy, two "Harpoons" can easily be carried by the "Hornet" in an inti-shipping configuration.

F/A-18 STORES OPTIONS
NOT TO SCALE

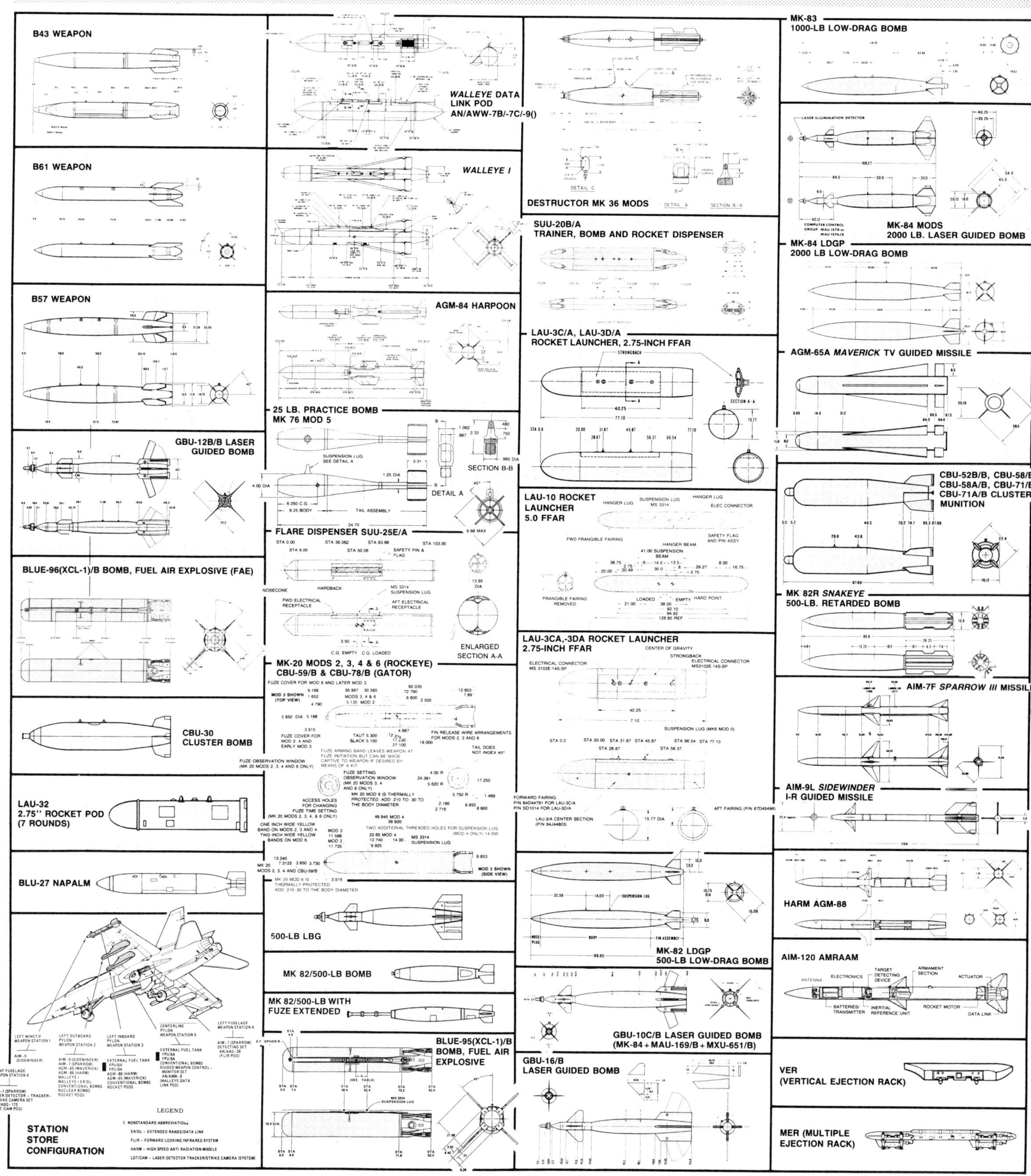

An early ''max payload'' configuration shows F/A-18A BuNo. 161248 equipped with two AIM-9Ls, four Mk.83 1,000 lb. bombs, three 330 gal. drop tanks (original oval configuration), an LST/SCAM pod on its starboard engine nacelle, and a FLIR pod on its port.

An extraordinary dense load of no less than 30 Mk.82 high-drag ''Snakeye'' bombs suspended from all four wing pylons and the centerline pylon. Additionally, AIM-9L ''Sidewinders'' are mounted on each wingtip rail.

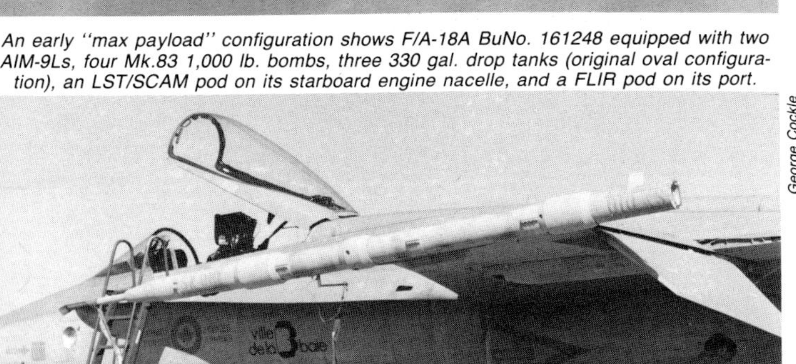

A Kelvin ACMI (Air Combat Maneuvering Instrumentation) pod attached to the port wingtip of a CAF CF-18A. The ACMI pod is a standardized evaluation device used to perfect air combat technique by recording ACM activity in training exercises.

A single AN/ALQ-167 METE (Multiple Environment Threat Emitter) pod attached to the centerline pylon of a VX-5 F/A-18A. This unit, which contains a AN/DLQ-3 transmitter, is used for electronic warfare training.

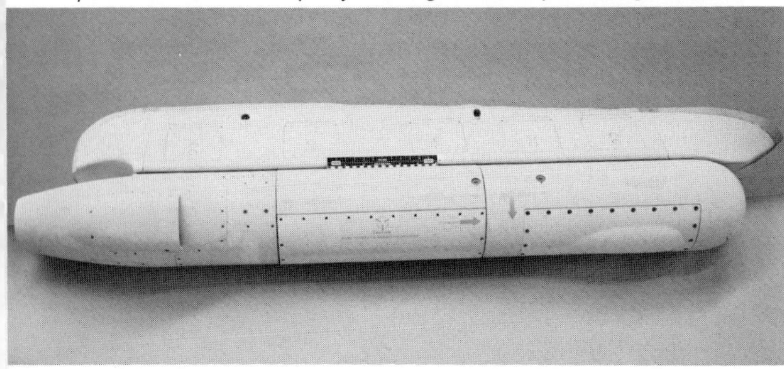

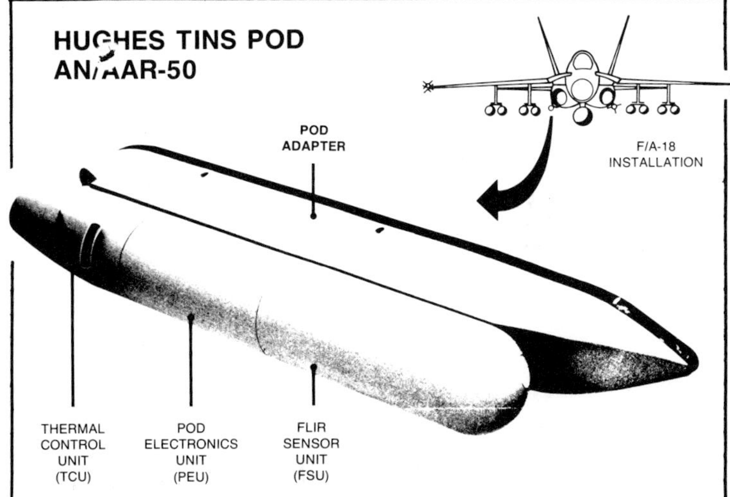

The Hughes AN/AAR-50 TINS (Thermal Imaging Navigation System) pod, when carried by the ''Hornet'', is attached to the starboard intake mounting station. TINS provides the pilot with infrared imagery suitable for low-light-level navigation technique.

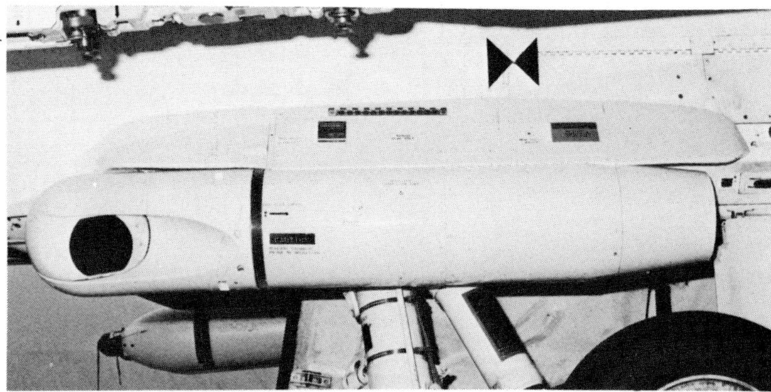

The Ford AN/AAS-38 FLIR pod is 72 in. long, 13 in. in diameter, and weighs 340 lbs. It has a designed-in reliability of 200 hours MTBF and comes equipped with fault isolation, built-in-test, and weapon replaceable assemblies to facilitate speed of repair.

Front view of 6,000 rpm General Electric M61A1 20mm rotary cannon and 570 round ammunition drum. This gun system sometimes is referred to as the GAU-11.

To avoid serious center-of-gravity shift problems as rounds are expended, the empty cartridges are retained in the nose of the aircraft and not ejected overboard. Unlike many of its predecessors, the M61A1 utilizes linkless feed. The six barrels rotate counter-clockwise (looking at them from the rear).

The M61A-1 has proven extremely reliable with only 1 failure per 10,000 rounds fired now noted. This dependability is due in part to the ammunition being fed to the firing chambers by external power, rather than being dependent upon the recoil action or gas generation of the weapon itself.

Aft end of the M61A-1 reveals gear drives, hydraulic umbilicals, and large cooling hose to channel cooling air into areas where greatest heat is generated.

A Canadian CF-18A test aircraft carries eight BL-755 cluster bomb units on its four wing pylons. BL-755 release tests were recorded with fuselage cameras.

One of the most recent "Hornet" developments is the Loral Defense Systems synthetic aperture radar pod which contains an AN/UPD-8 equipped with a data processor and data link. The latter transmits real-time digitized imagery to ground stations. The pod weighs less than 1,500 lbs., is ram air cooled, and requires 2.6 kw of power.

The AN/ASQ-173 laser spot tracker/strike camera (LST/CAM) pod is used when the "Hornet" is configured for ground attack missions. The pod bolts to the starboard engine nacelle station in place of a single AIM-7F. It enables cooperative precision bombing of laser designated targets (i.e., several aircraft equipped with this system can coordinate their targeting and thus utilize weapons more judiciously). Additionally, it provides damage assessment imagery for post flight review.